Problem-Solving and Selected Topics in Euclidean Geometry

Sotirios E. Louridas • Michael Th. Rassias

Problem-Solving and Selected Topics in Euclidean Geometry

In the Spirit of the Mathematical Olympiads

Foreword by Michael H. Freedman

Sotirios E. Louridas
Athens, Greece

Michael Th. Rassias
Department of Mathematics
ETH Zurich
Zurich, Switzerland

ISBN 978-1-4899-8678-8 ISBN 978-1-4614-7273-5 (eBook)
DOI 10.1007/978-1-4614-7273-5
Springer New York Heidelberg Dordrecht London

Mathematics Subject Classification: 51-XX, 51-01

Softcover re-print of the Hardcover 1st edition 2013

Printed on acid-free paper

Springer is part of Springer Science+Business Media (www.springer.com)

Foreword

Geometry has apparently fallen on hard times. I learned from this excellent treatise on plane geometry that U.S. President James A. Garfield constructed his own proof of the Pythagorean Theorem in 1876, four years before being elected to an unfortunately brief presidency.

In a recent lecture, Scott Aaronson (MIT) offered a tongue-in-cheek answer to the question: "Suppose there is a short proof that $P \neq NP$?" with, "Suppose space aliens assassinated President Kennedy to prevent him from discovering such a proof?" I found it pleasant to wonder which half of the two clauses was *less* probable. Sadly he concluded that it was more likely that space aliens were behind Kennedy's assassination than that a modern president would be doing mathematics. Perhaps this book offers hope that what was possible once will be possible again.

Young people need such texts, grounded in our shared intellectual history and challenging them to excel and create a continuity with the past. Geometry has seemed destined to give way in our modern computerized world to algebra. As with Michael Th. Rassias' previous homonymous book on number theory, it is a pleasure to see the mental discipline of the ancient Greeks so well represented to a youthful audience.

Michael H. Freedman

Microsoft Station Q
CNSI Bldg., Office 2245
University of California
Santa Barbara, CA 93106-6105
USA

Foreword

Acknowledgements

We feel deeply honored and grateful to Professor Michael H. Freedman, who has written the Foreword of the book.

We would like to express our thanks to Professors D. Andrica, M. Bencze, S. Markatis, and N. Minculete for reading the manuscript and providing valuable suggestions and comments which have helped to improve the presentation of the book. We would also like to thank Dr. A. Magkos for his useful remarks.

We would like to express our appreciation to the Hellenic Mathematical Society for the excellent program of preparation of the young contestants for their participation in Mathematical Olympiads.

Last but not least we would like to offer our thanks to Dr. F.I. Travlopanos for his invaluable help and for his useful suggestions.

Finally, it is our pleasure to acknowledge the superb assistance provided by the staff of Springer for the publication of the book.

Sotirios E. Louridas
Michael Th. Rassias

Acknowledgements

Contents

Chapter 1
Introduction

μη μου τους κύκλους τάραττε.
(Do not disturb my circles.)
Archimedes (287 BC–212 BC)

In this chapter, we shall present an overview of Euclidean Geometry in a general, non-technical context.

1.1 The Origin of Geometry

Generally, we could describe geometry as the mathematical study of the physical world that surrounds us, if we consider it to extend indefinitely. More specifically, we could define geometry as the mathematical investigation of the measure, the properties and relationships of points, lines, angles, shapes, surfaces, and solids.

It is commonly accepted that basic methods of geometry were first discovered and used in everyday life by the Egyptians and the Babylonians. It is remarkable that they could calculate simple areas and volumes and they had closely approximated the value of π (the ratio of the circumference to the diameter of a circle).

However, even though the Egyptians and the Babylonians had undoubtedly mastered some geometrical techniques, they had not formed a mathematical system of geometry as a theoretical science comprising definitions, theorems, and proofs. This was initiated by the Greeks, approximately during the seventh century BC.

It is easy to intuitively understand the origin of the term geometry, if we etymologically study the meaning of the term. The word *geometry* originates from the Greek word γεωμετρία, which is formed by two other Greek words: The word γη, which means *earth* and the word μέτρον, which means *measure*. Hence, geometry actually means the *measurement of the earth*, and originally, that is exactly what it was before the Greeks. For example, in approximately 240 BC, the Greek mathematician Eratosthenes used basic but ingenious methods of geometry that were developed theoretically by several Greek mathematicians before his time in order to measure the Earth's circumference. It is worth mentioning that he succeeded to do so, with an error of less than 2 % in comparison to the exact length of the circumference as we know it today. Therefore, it is evident that geometry arose from practical activity.

S.E. Louridas, M.Th. Rassias, *Problem-Solving and Selected Topics in Euclidean Geometry*, DOI 10.1007/978-1-4614-7273-5_1,

Geometry was developed gradually as an abstract theoretical science by mathematicians/philosophers, such as Thales, Pythagoras, Plato, Apollonius, Euclid, and others. More specifically, Thales, apart from his intercept theorem, is also the first mathematician to whom the concept of proof by induction is attributed. Moreover, Pythagoras created a school known as the Pythagoreans, who discovered numerous theorems in geometry. Pythagoras is said to be the first to have provided a deductive proof of what is known as the Pythagorean Theorem.

Theorem 1.1 (Pythagorean Theorem) *In any right triangle with sides of lengths a, b, c, where c is the length of the hypotenuse, it holds*

$$a^2 + b^2 = c^2. \tag{1.1}$$

The above theorem has captured the interest of both geometers and number theorists for thousands of years. Hundreds of proofs have been presented since the time of Pythagoras. It is amusing to mention that even the 20th president of the United States, J.A. Garfield, was so much interested in this theorem that he managed to discover a proof of his own, in 1876.

The number theoretic aspect of the Pythagorean Theorem is the study of the integer values a, b, c, which satisfy Eq. (1.1). Such triples of integers (a, b, c) are called *Pythagorean triples* [86]. Mathematicians showed a great interest in such properties of integers and were eventually lead to the investigation of the solvability of equations of the form

$$a^n + b^n = c^n,$$

where $a, b, c \in \mathbb{Z}^+$ and $n \in \mathbb{N}, n > 2$.

These studies lead after hundreds of years to Wiles' celebrated proof of Fermat's Last Theorem [99], in 1995.

Theorem 1.2 (Fermat's Last Theorem) *It holds*

$$a^n + b^n \neq c^n,$$

for every $a, b, c \in \mathbb{Z}^+$ and $n \in \mathbb{N}, n > 2$.

Let us now go back to the origins of geometry. The first rigorous foundation which made this discipline a well-formed mathematical system was provided in Euclid's *Elements* in approximately 300 BC.

The Elements are such a unique mathematical treatise that there was no need for any kind of additions or modifications for more than 2000 years, until the time of the great Russian mathematician N.I. Lobačevskiĭ (1792–1856) who developed a new type of geometry, known as hyperbolic geometry, in which Euclid's parallel postulate was not considered.

1.2 A Few Words About Euclid's Elements

For more than 2000 years, the Elements had been the absolute point of reference of deductive mathematical reasoning. The text itself and some adaptations of it by great mathematicians, such as A.M. Legendre (1752–1833) and J. Hadamard (1865–1963), attracted a lot of charismatic minds to Mathematics. It suffices to recall that it was the lecture of Legendre's Elements that attracted E. Galois (1811–1832), one of the greatest algebraists of all time, to Mathematics.

Euclid's Elements comprise 13 volumes that Euclid himself composed in Alexandria in about 300 BC. More specifically, the first four volumes deal with figures, such as triangles, circles, and quadrilaterals. The fifth and sixth volumes study topics such as similar figures. The next three volumes deal with a primary form of elementary number theory, and the rest study topics related to geometry. It is believed that the Elements founded logic and modern science.

In the Elements, Euclid presented some assertions called *axioms*, which he considered to be a set of self-evident premises on which he would base his mathematical system. Apart from the axioms, Euclid presented five additional assertions called *postulates*, whose validity seemed less certain than the axioms', but still considered to be self-evident.

The Axioms

1. Things that are equal to the same thing are also equal to one another.
2. If equals are to be added to equals, then the wholes will be equal.
3. If equals are to be subtracted from equals, then the remainders will be equal.
4. Things that coincide to one another are equal to one another.
5. The whole is greater than the part.

The Postulates

1. There is a unique straight line segment connecting two points.
2. Any straight line segment can be indefinitely extended (continuously) in a straight line.
3. There exists a circle with any center and any value for its radius.
4. All right angles are equal to one another.
5. If a straight line intersects two other straight lines, in such a way that the sum of the inner angles on the same side is less than two right angles, then the two straight lines will eventually meet if extended indefinitely.

Regarding the first four postulates of Euclid, the eminent mathematical physicist R. Penrose (1931–) in his book *The Road to Reality—A Complete Guide to the Laws of the Universe*, Jonathan Cape, London, 2004, writes:

> Although Euclid's way of looking at geometry was rather different from the way that we look at it today, his first four postulates basically encapsulated our present-day notion of a (two-dimensional) metric space with complete homogeneity and isotropy, and infinite in extent. In fact, such a picture seems to be in close accordance with the very large-scale spatial nature of the actual universe, according to modern cosmology.

The fifth postulate, known as the *parallel postulate*, has drawn a lot of attention since Euclid's time. This is due to the fact that the parallel postulate does not seem to be self-evident. Thus, a lot of mathematicians over the centuries have tried to provide a proof for it, by the use of the first four postulates. Even though several proofs have been presented, sooner or later a mistake was discovered in each and every one of them. The reason for this was that all the proofs were at some point making use of some statement which seemed to be obvious or self-evident but later turned out to be equivalent to the parallel postulate itself. The independence of the parallel postulate from Euclid's other axioms was settled in 1868 by Eugenio Beltrami (1836–1900).

The close examination of Euclid's axiomatics from the formalistic point of view culminated at the outset of the twentieth century, in the seminal work of David Hilbert (1862–1943), which influenced much of the subsequent work in Mathematics.

However, to see the Elements as an incomplete formalist foundation-building for the Mathematics of their time is only an a posteriori partial view. Surely, a full of respect mortal epigram, but not a convincing explanation for the fact that they are a permanent source of new inspiration, both in foundational research and in that on working Mathematics.

It is no accident that one of the major mathematicians of the twentieth century, G.H. Hardy, in his celebrated *A Mathematician's Apology* takes his two examples of important Mathematics that will always be "fresh" and "significant" from the Elements. Additionally, the eminent logician and combinatorist D. Tamari (1911–2006) insisted on the fact that Euclid was the first thinker to expose a well-organized scientific theory without the mention or use of extra-logical factors. Thus, according to D. Tamari, Euclid must be considered as the founder of modern way of seeing scientific matters. References [1–99] provide a large amount of theory and several problems in Euclidean Geometry and its applications.

Chapter 2
Preliminaries

Where there is matter, there is geometry.
Johannes Kepler (1571–1630)

2.1 Logic

2.1.1 Basic Concepts of Logic

Let us consider A to be a non-empty set of mathematical objects. One may construct various expressions using these objects. An expression is called a *proposition* if it can be characterized as "true" or "false."

Example

1. "The number $\sqrt{2}$ is irrational" is a true proposition.
2. "An isosceles triangle has all three sides mutually unequal" is a false proposition.
3. "The median and the altitude of an equilateral triangle have different lengths" is false.
4. "The diagonals of a parallelogram intersect at their midpoints" is true.

A proposition is called *compound* if it is the juxtaposition of propositions connected to one another by means of *logical connectives*. The truth values of compound propositions are determined by the truth values of their constituting propositions and by the behavior of logical connectives involved in the expression. The set of propositions equipped with the operations defined by the logical connectives becomes the *algebra of propositions*. Therefore, it is important to understand the behavior of logical connectives.

The logical connectives used in the algebra of propositions are the following:

$$\wedge(\text{and}) \qquad \vee(\text{or}) \qquad \Rightarrow (\text{if}\ldots\text{then})$$

$$\Leftrightarrow (\text{if and only if}) \quad \text{and} \quad \neg(\text{not}).$$

The mathematical behavior of the connectives is described in the *truth tables*, seen in Tables 2.1, 2.2, 2.3, 2.4, and 2.5.

S.E. Louridas, M.Th. Rassias, *Problem-Solving and Selected Topics in Euclidean Geometry*, DOI 10.1007/978-1-4614-7273-5_2,

Table 2.1 Truth table for $\wedge$

a	b	$a \wedge b$
T	T	T
T	F	F
F	T	F
F	F	F

Table 2.2 Truth table for $\vee$

a	b	$a \vee b$
T	T	T
T	F	T
F	T	T
F	F	F

Table 2.3 Truth table for $\Rightarrow$

a	b	$a \Rightarrow b$
T	T	T
T	F	F
F	T	T
F	F	T

Table 2.4 Truth table for $\Leftrightarrow$

a	b	$a \Leftrightarrow b$
T	T	T
T	F	F
F	T	F
F	F	T

Table 2.5 Truth table for $\neg$

a	$\neg a$
T	F
F	T

In the case when

$$a \Rightarrow b \quad \text{and} \quad b \Rightarrow a \tag{2.1}$$

are simultaneously true, we say that a and b are "equivalent" or that "a if and only if b" or that a is a necessary and sufficient condition for b.

Let us now focus on mathematical problems. A mathematical problem is made up from the hypothesis and the conclusion. The hypothesis is a proposition assumed to be true in the context of the problem. The conclusion is a proposition whose truth one is asked to show. Finally, the solution consists of a sequence of logical implications

$$a \Rightarrow b \Rightarrow c \Rightarrow \cdots . \tag{2.2}$$

Mathematical propositions are categorized in the following way: *Axioms, theorems, corollaries, problems*.

Axioms are propositions considered to be true without requiring a proof. Another class of propositions are the *lemmata*, which are auxiliary propositions; the proof of a lemma is a step in the proof of a theorem.

In Euclidean Geometry, we have three basic axioms concerning comparison of figures:

1. Two figures, A and B, are said to be *congruent* if and only if there exists a translation, or a rotation, or a symmetry, or a composition of these transformations such that the image of figure A coincides with figure B.
2. Two figures which are congruent to a third figure are congruent to each other.
3. A part of a figure is a subset of the entire figure.

2.1.2 On Related Propositions

Consider the proposition

$$p : a \Rightarrow b.$$

Then:

1. The *converse* of proposition p is the proposition

$$q : b \Rightarrow a. \tag{2.3}$$

2. The *inverse* of proposition p is the proposition

$$r : \neg a \Rightarrow \neg b. \tag{2.4}$$

3. The *contrapositive* of proposition p is the proposition

$$s : \neg b \Rightarrow \neg a. \tag{2.5}$$

Example Consider the proposition

p: *If a convex quadrilateral is a parallelogram then its diagonals bisect each other.*

1. The *converse* proposition of p is q:
   If the diagonals of a converse quadrilateral bisect each other, then it is a parallelogram.

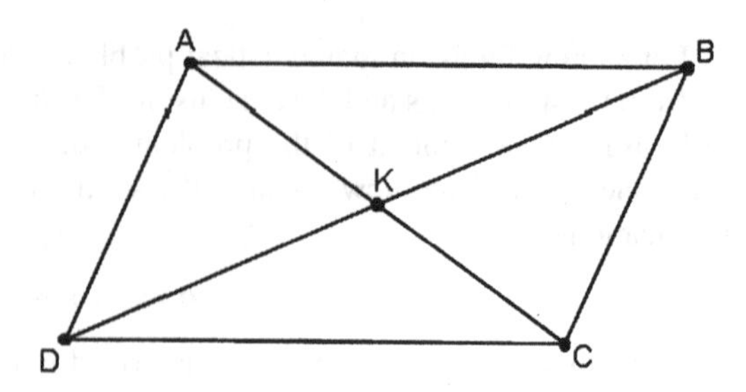

Fig. 2.1 Example 2.1.1

2. The *inverse* proposition of p is r:
   If a convex quadrilateral is not a parallelogram, then its diagonals do not bisect each other.
3. The *contrapositive* proposition of p is s:
   If the diagonals of a convex quadrilateral do not bisect each other, then it is not a parallelogram.

2.1.3 On Necessary and Sufficient Conditions

Proofs of propositions are based on proofs of the type

$$a \Rightarrow b,$$

where a is the set of hypotheses and b the set of conclusions. In this setup, we say that condition a is sufficient for b and that b is necessary for a. Similarly, in the case of the converse proposition

$$q : b \Rightarrow a, \tag{2.6}$$

condition b is sufficient for a and condition a is necessary for b.

In the case where both

$$a \Rightarrow b \quad \text{and} \quad b \Rightarrow a \tag{2.7}$$

are true, we have

$$a \Leftrightarrow b, \tag{2.8}$$

which means that a is a necessary and sufficient condition for b.

Example 2.1.1 A necessary and sufficient condition for a convex quadrilateral to be a parallelogram is that its diagonals bisect.

Proof Firstly, we assume that the quadrilateral $ABCD$ is a parallelogram (see Fig. 2.1). Let K be the point of intersection of its diagonals. We use the property that the opposite sides of a parallelogram are parallel and equal and we have that

$$AB = DC \tag{2.9}$$

and

$$\widehat{KAB} = \widehat{KCD} \tag{2.10}$$

since the last pair of angles are alternate interior. Also, we have

$$\widehat{ABK} = \widehat{CDK} \tag{2.11}$$

as alternate interior angles. Therefore, the triangles KAB and KDC are equal, and hence

$$KB = DK \quad \text{and} \quad AK = KC. \tag{2.12}$$

For the converse, we assume now that the diagonals of a convex quadrilateral $ABCD$ bisect each other. Then, if K is their intersection, we have that

$$KA = CK \quad \text{and} \quad KB = DK. \tag{2.13}$$

Furthermore, we have

$$\widehat{BKA} = \widehat{DKC} \tag{2.14}$$

because they are corresponding angles. Hence, the triangles KAB and KDC are equal. We conclude that

$$AB = DC \tag{2.15}$$

and also that

$$AB \parallel DC \tag{2.16}$$

since

$$\widehat{KAB} = \widehat{KCD}. \tag{2.17}$$

Therefore, the quadrilateral $ABCD$ is a parallelogram. □

2.2 Methods of Proof

We now present the fundamental methods used in geometric proofs.

2.2.1 Proof by Analysis

Suppose we need to show that

$$a \Rightarrow b. \tag{2.18}$$

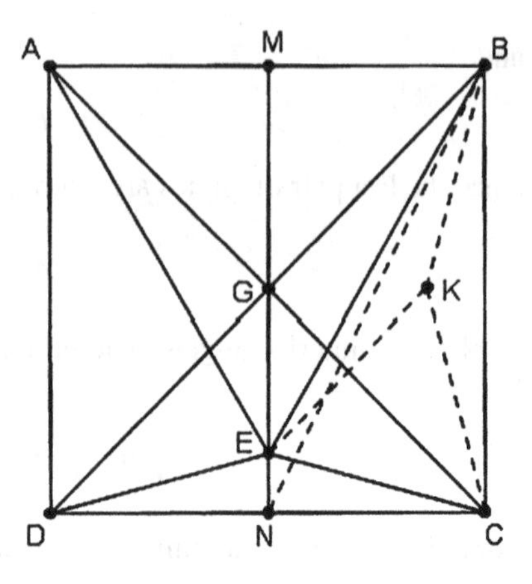

Fig. 2.2 Proof by analysis (Example 2.2.1)

We first find a condition b_1 whose truth guarantees the truth of b, i.e., a sufficient condition for b. Subsequently, we find a condition b_2 which is sufficient for b_1. Going *backwards* in this way, we construct a chain of conditions

$$b_n \Rightarrow b_{n-1} \Rightarrow \cdots \Rightarrow b_1 \Rightarrow b,$$

with the property that b_n is true by virtue of a being true. This completes the proof.

Example 2.2.1 Consider the square $ABCD$. From the vertices C and D we consider the half-lines that intersect in the interior of $ABCD$ at the point E and such that

$$\widehat{CDE} = \widehat{ECD} = 15°.$$

Show that the triangle EAB is equilateral (see Fig. 2.2).

Proof We observe that

$$AD = BC, \tag{2.19}$$

since they are sides of a square. Furthermore, we have (see Fig. 2.2)

$$\widehat{CDE} = \widehat{ECD} = 15° \tag{2.20}$$

hence

$$\widehat{EDA} = \widehat{BCE} = 75°$$

and

$$ED = EC, \tag{2.21}$$

since the triangle EDC is isosceles. Therefore, the triangles ADE and BEC are equal and thus

$$EA = EB. \tag{2.22}$$

Therefore, the triangle EAB is isosceles. In order to show that the triangle EAB is, in fact, equilateral, it is enough to prove that

$$EB = BC = AB. \tag{2.23}$$

In other words, it is sufficient to show the existence of a point K such that the triangles KBE and KCB are equal. In order to use our hypotheses, we can choose K in such a way that the triangles KBC and EDC are equal. This will work as long as K is an interior point of the square.

Let G be the center of the square $ABCD$. Then, if we consider a point K such that the triangles KCB and EDC are equal, we have the following:

$$GN < GB, \tag{2.24}$$

and hence

$$\widehat{GBN} < \widehat{GNB} = \widehat{NBC}. \tag{2.25}$$

Therefore,

$$2\widehat{GBN} < 45^\circ, \tag{2.26}$$

so

$$\widehat{GBN} < 22.5^\circ, \tag{2.27}$$

and hence

$$\widehat{NBC} > 22.5^\circ. \tag{2.28}$$

Therefore,

$$\widehat{NBC} > \widehat{KBC}, \tag{2.29}$$

where M, N are the midpoints of the sides AB, DC, respectively. Therefore, the point K lies in the interior of the angle $\widehat{EBC}$. We observe that

$$\widehat{KCE} = 90^\circ - 15^\circ - 15^\circ = 60^\circ, \tag{2.30}$$

with

$$KC = CE. \tag{2.31}$$

Therefore, the isosceles triangle CKE has an angle of 60° and thus it is equilateral, implying that

$$EK = KC = CE \tag{2.32}$$

and

$$\widehat{BKE} = 360^\circ - 60^\circ - 150^\circ = 150^\circ. \tag{2.33}$$

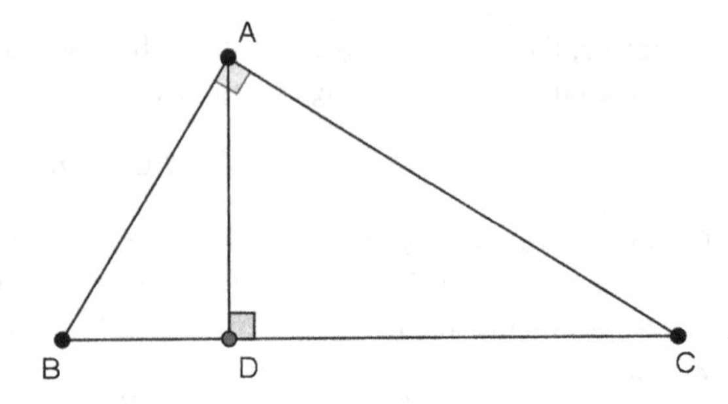

Fig. 2.3 Proof by synthesis (Example 2.2.2)

Therefore, the triangles KCB and KBE are equal, and hence

$$EB = BC. \tag{2.34}$$

□

2.2.2 *Proof by Synthesis*

Suppose we need to show that

$$a \Rightarrow b. \tag{2.35}$$

The method we are going to follow consists of combining proposition a with a number of suitable true propositions and creating a sequence of necessary conditions leading to b.

Example 2.2.2 Let ABC be a right triangle with $\widehat{BAC} = 90°$ and let AD be the corresponding height. Show that

$$\frac{1}{AD^2} = \frac{1}{AB^2} + \frac{1}{AC^2}.$$

Proof First, consider triangles ABD and CAB (see Fig. 2.3). Since they are both right triangles and

$$\widehat{BAD} = \widehat{ACB},$$

they are similar. Then we have (see Fig. 2.3)

$$\frac{AB}{BD} = \frac{BC}{AB},$$

and therefore,

$$AB^2 = BD \cdot BC.$$

Similarly,

$$AC^2 = DC \cdot BC$$

and

$$AD^2 = BD \cdot DC.$$

Hence

$$\begin{aligned}\frac{1}{AB^2} + \frac{1}{AC^2} &= \frac{1}{BD \cdot BC} + \frac{1}{DC \cdot BC} \\ &= \frac{BC}{BD \cdot BC \cdot DC} \\ &= \frac{1}{AD^2}.\end{aligned}$$

□

2.2.3 Proof by Contradiction

Suppose that we need to show

$$a \Rightarrow b. \tag{2.36}$$

We assume that the negation of proposition $a \Rightarrow b$ is true. Observe that

$$\neg(a \Rightarrow b) = a \wedge (\neg b). \tag{2.37}$$

In other words, we assume that given a, proposition b does not hold. If with this assumption we reach a false proposition, then we have established that

$$a \Rightarrow b \tag{2.38}$$

is true.

Example 2.2.3 Let ABC be a triangle and let D, E, Z be three points in its interior such that

$$3S_{DBC} < S_{ABC}, \tag{2.39}$$

$$3S_{EAC} < S_{ABC}, \tag{2.40}$$

$$3S_{ZAB} < S_{ABC}, \tag{2.41}$$

where S_{ABC} denotes the area of the triangle ABC and so on. Prove that the points D, E, Z cannot coincide.

Proof Suppose that inequalities (2.39), (2.40), and (2.41) hold true and let P be the point where D, E, Z coincide, that is,

$$D \equiv E \equiv Z \equiv P.$$

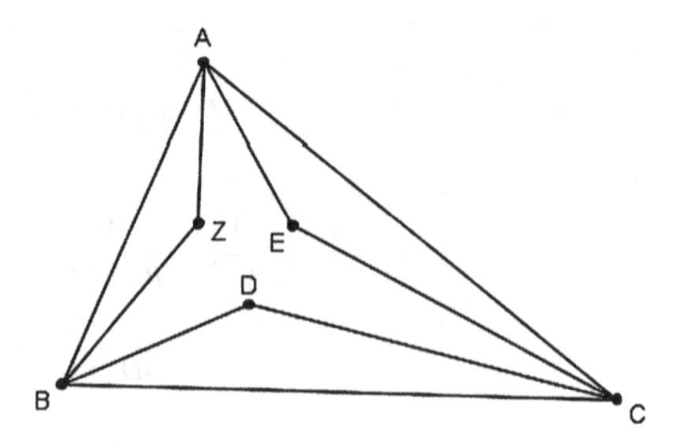

Fig. 2.4 Proof by contradiction (Example 2.2.3)

Then (see Fig. 2.4),

$$3[S_{PBC} + S_{PAC} + S_{PAB}] < 3S_{ABC} \tag{2.42}$$

and thus

$$S_{ABC} < S_{ABC}, \tag{2.43}$$

which is a contradiction. Therefore, when relations (2.39), (2.40), and (2.41) are satisfied, the three points D, E, Z cannot coincide. □

Now, we consider Example 2.2.1 from a different point of view.

Example 2.2.4 Let $ABCD$ be a square. From the vertices C and D we consider the half-lines that intersect in the interior of $ABCD$ at the point E and such that

$$\widehat{CDE} = \widehat{ECD} = 15°.$$

Show that the triangle EBA is equilateral.

Proof We first note that the triangle EBA is isosceles. Indeed, since by assumption

$$\widehat{CDE} = \widehat{ECD} = 15°$$

one has

$$\widehat{EDA} = \widehat{BCE} = 75° \Rightarrow ED = EC. \tag{2.44}$$

We also have that $AD = BC$, and therefore the triangles ADE and BEC are equal and thus $EA = EB$, which means that the point E belongs to the common perpendicular bisector MN of the sides AB, DC of the square $ABCD$.

Let us assume that the triangle EBA is not equilateral. Then, there exists a point Z on the straight line segment MN, Z different from E, such that the triangle ZAB is equilateral. Indeed, by choosing the point Z on the half straight line MN such that

$$MZ = \frac{AB\sqrt{3}}{2} < AB = AD = MN,$$

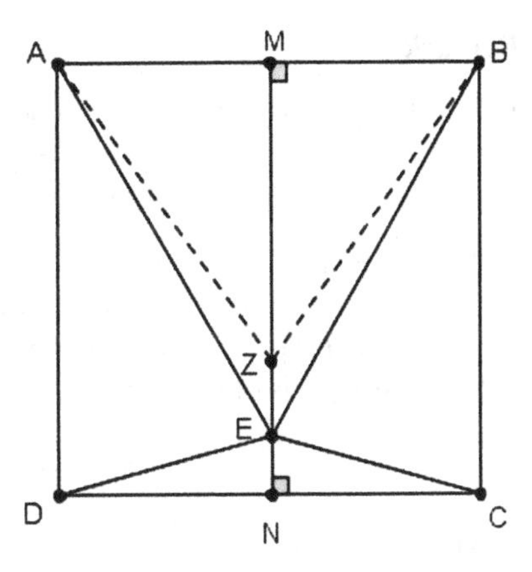

Fig. 2.5 Proof by contradiction (Example 2.2.4)

the point Z is an interior point of the straight line segment MN and the equilaterality of the triangle ZBA shall be an obvious consequence (see Fig. 2.5).

We observe that

$$\widehat{DAZ} = \widehat{ZBC} = 30^\circ \quad \text{and} \quad DA = ZA = AB = BZ = BC.$$

Thus

$$2\widehat{ZDA} = 180^\circ - 30^\circ \Rightarrow \widehat{ZDA} = 75^\circ,$$

which implies that

$$\widehat{CDZ} = 90^\circ - 75^\circ = 15^\circ. \tag{2.45}$$

Hence the points E, Z coincide, which is a contradiction. Therefore, the triangle EBA is equilateral. □

2.2.4 *Mathematical Induction*

This is a method that can be applied to propositions which depend on natural numbers. In other words, propositions of the form

$$p(n), \quad n \in \mathbb{N}. \tag{2.46}$$

The proof of proposition (2.46) is given in three steps.

1. One shows that $p(1)$ is true.
2. One assumes proposition $p(n)$ to be true.
3. One shows that $p(n+1)$ is true.

Remarks

- If instead of proposition (2.46) one needs to verify the proposition

$$p(n), \quad \forall n \in \mathbb{N} \setminus \{1, 2, \ldots, N\}, \tag{2.47}$$

then the first step of the process is modified as follows:

Instead of showing $p(1)$ to be true, one shows $p(N+1)$ to be true. After that, we assume proposition $p(n)$ to be true and we prove that $p(n+1)$ is true.

- Suppose $p(n)$ is of the form

$$p(n): \quad k(n) \geq q(n), \quad \forall n \geq N, \tag{2.48}$$

where $n, N \in \mathbb{N}$.

Suppose that we have proved

$$k(N) = q(N). \tag{2.49}$$

We must examine the existence of at least one natural number $m > N$ for which $k(m) > q(m)$.

This is demonstrated in the following example.

Example 2.2.5 Let ABC be a right triangle with $\widehat{BAC} = 90°$, with lengths of sides $BC = a$, $AC = b$, and $AB = c$. Prove that

$$a^n \geq b^n + c^n, \quad \forall n \in \mathbb{N}\backslash\{1\}. \tag{2.50}$$

Proof Applying the induction method we have.

- Evidently, for $n = 2$, the Pythagorean Theorem states that Eq. (2.50) holds true and is, in fact, an equality. We shall see that for $n = 3$ it holds

$$a^3 > b^3 + c^3.$$

In order to prove this, it suffices to show

$$a\left(b^2 + c^2\right) > b^3 + c^3. \tag{2.51}$$

To show (2.51), it is enough to show

$$ab^2 + ac^2 - b^3 - c^3 > 0, \tag{2.52}$$

for which it is sufficient to show

$$b^2(a - b) + c^2(a - c) > 0. \tag{2.53}$$

This inequality holds because the left hand is strictly positive, since $a > c$ and $a > b$.

- We assume that

$$a^n > b^n + c^n \tag{2.54}$$

for $n \in \mathbb{N} \setminus \{1, 2\}$.
- We shall prove that

$$a^{n+1} > b^{n+1} + c^{n+1}. \tag{2.55}$$

For (2.55) it suffices to show that

$$a\big(b^n + c^n\big) - b^{n+1} - c^{n+1} > 0, \tag{2.56}$$

for which, in turn, it is enough to show that

$$b^n(a - b) + c^n(a - c) > 0. \tag{2.57}$$

Again, in the last inequality the left hand side term is greater than 0, since $a > c$ and $a > b$. Therefore, (2.50) is true. □

Chapter 3
Fundamentals on Geometric Transformations

Geometry is knowledge of the eternally existent.
Pythagoras (570 BC–495 BC)

A topic of high interest for problem-solving in Euclidean Geometry is the determination of a point by the use of geometric transformations: translation, symmetry, homothety, and inversion. The knowledge of geometric transformations allows us to understand the geometric behavior of plane figures produced by them.

3.1 A Few Facts

1. In order to create a geometric figure, it is enough to have a point and a clear mathematical way (see Fig. 3.1) in which the point moves on the plane in order to produce this shape. We can then say that the point traverses the planar shape.
2. Two points of the plane that move in the plane in the same mathematical way traverse the same or equal planar shapes.
3. A bijective correspondence (bijective mapping) is defined between two shapes if there is a law that to each point of the one shape corresponds one and only one point of the other shape, and conversely. The shapes are then called *corresponding*.

Between two equal shapes, there is always a bijective correspondence. The converse is not always true. For example, see Fig. 3.2. It is clear that the easiest way to establish a bijective transformation between the straight line segment AB and the crooked line segment KLM, where A, B are the projections of K, M on AB, respectively, is to consider the projection of every point of KLM to the corresponding point of AB.

The projection is unique because from every point of the plane there exists a unique line perpendicular to AB.

If $AB \parallel l$, then the semi-circumference with center O and diameter AB, with A, B excluded, can be matched to l in the following way: From the point O, we consider half-lines Ox (see Fig. 3.3 and Fig. 3.4).

Then to each intersection point M of Ox with the semi-circumference, one can select the unique point N which is the intersection of l with Ox, and conversely.

S.E. Louridas, M.Th. Rassias, *Problem-Solving and Selected Topics in Euclidean Geometry*, DOI 10.1007/978-1-4614-7273-5_3,

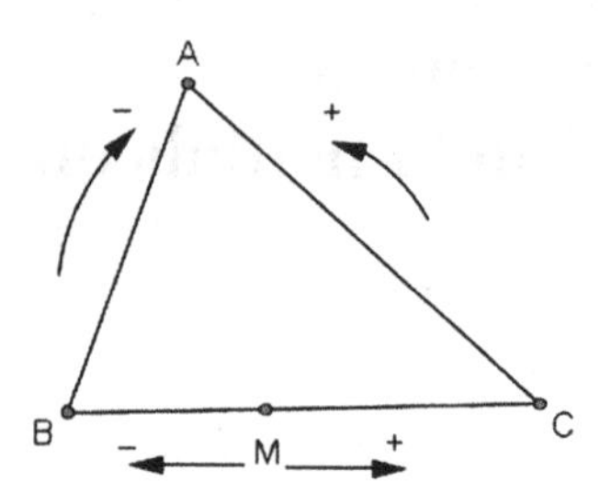

Fig. 3.1 A few facts (Sect. 3.1)

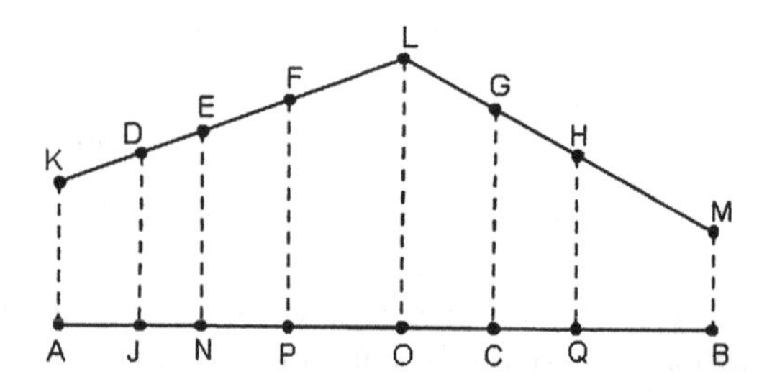

Fig. 3.2 A few facts (Sect. 3.1)

Observation We shall say that two corresponding shapes are *traversed similarly* if during their traversal their points are traversed in the same order.

- A circumference
  - (i) is traversed in the *positive* direction if a "moving" point traverses it anticlockwise.
  - (ii) is traversed in the *negative* direction if a "moving" point traverses it clockwise. We think of the clock as lying on the same plane as our shape (see Fig. 3.1).

Remark Assume that we have two congruent shapes. Then we can construct a correspondence between their points in the following way:

$$A_1A_2\dots A_n = B_1B_2\dots B_n, \tag{3.1}$$

with

$$\begin{aligned} A_1A_2 &= B_1B_2, \\ A_2A_3 &= B_2B_3, \\ &\dots \\ A_{n-1}A_n &= B_{n-1}B_n, \\ A_nA_1 &= B_nB_1, \end{aligned}$$

and assume that from their vertices A_1, B_1 two points start moving at the same speed, traversing their respective circumferences. Then at a given point in time t_0 the points have covered equal paths.

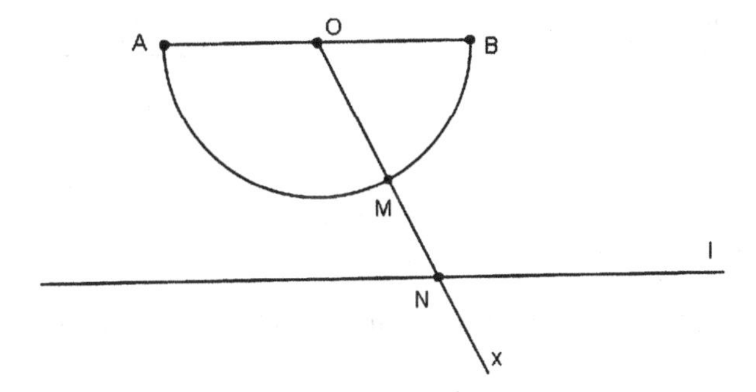

Fig. 3.3 A few facts (Sect. 3.1)

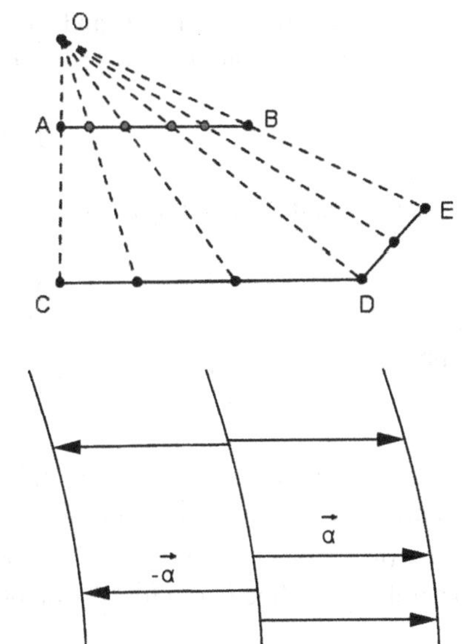

Fig. 3.4 A few facts (Sect. 3.1)

Fig. 3.5 Translation (Sect. 3.2)

3.2 Translation

1. Let A be a point and let $\overrightarrow{a}$ be a vector. Consider the vector $\overrightarrow{AB} = \overrightarrow{a}$. Then the point B is the translation of A by the vector $\overrightarrow{a}$.
2. Let S be a shape and let $\overrightarrow{a}$ be a vector. The translation S_1 of the shape S is the shape whose points are the translations of the points of S by the vector $\overrightarrow{a}$. If S_1 is the translation of S by $\overrightarrow{a}$, then S_2 is the opposite translation of S if S_2 is the translation of S by $-\overrightarrow{a}$ (see Fig. 3.5).
3. The translation of a shape gives a shape equal to the initial shape.

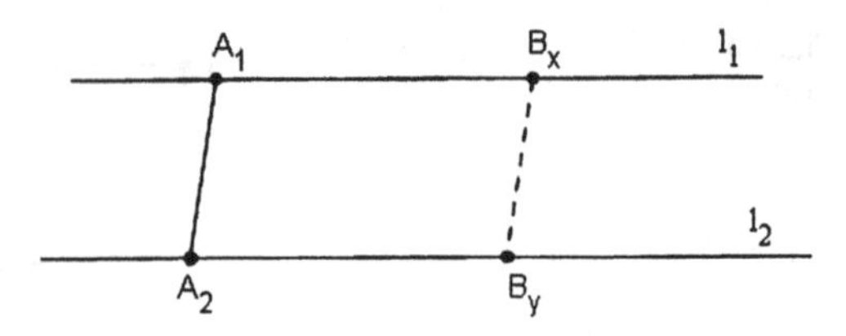

Fig. 3.6 Translation (Sect. 3.2)

4. If l_1 and l_2 are parallel lines, then these lines are translations of each other (see Fig. 3.6).

Proof Let A_1 be a point on l_1 and let A_2 be a point on l_2. Consider the vector $\overrightarrow{A_1A_2}$. Let B_x be a point on l_1. We consider B_y to be a point such that

$$\overrightarrow{B_xB_y} = \overrightarrow{A_1A_2}. \tag{3.2}$$

This implies that the point B_y lies on l_2, since

$$\overrightarrow{A_1A_2} = \overrightarrow{B_xB_y}$$

implies

$$A_2B_y \parallel A_1B_x.$$

Now, from the well-known axiom of Euclid (see Introduction), there exists only one parallel line to l_1 that passes through A_2, and that line is l_2. The point B_y is a unique point of l_2, corresponding to B_x with respect to the translation by the vector $\overrightarrow{A_1A_2}$. Therefore, the line l_2 is a translation of the line l_1. □

Definition 3.1 Two translations are said to be *consecutive* if the first one translates the shape S to the shape S_1 and the second translates S_1 to the shape S_2.

Theorem 3.1 *Two consecutive translations defined by vectors of different directions can be replaced by one translation, which is defined by one vector which is the vector sum of the other two vectors.*

Proof We consider the vectors $\overrightarrow{OA_1}$ and $\overrightarrow{OA_2}$ that have a common starting point O and are equal to the vectors defining the translation (see Fig. 3.7). Let M be a point of the initial shape. After the first translation, M is translated to M_1 so that

$$\overrightarrow{MM_1} = \overrightarrow{OA_1}. \tag{3.3}$$

After the second translation, M_1 is translated to the point M_2' so that

$$\overrightarrow{M_1M_2'} = \overrightarrow{OA_2}. \tag{3.4}$$

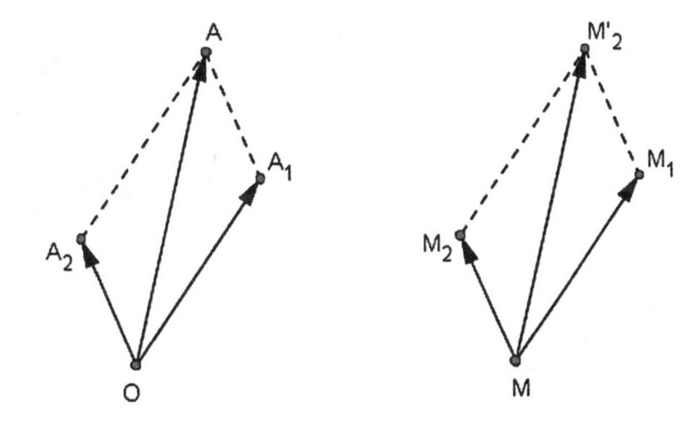

Fig. 3.7 Translation (Sect. 3.2)

Therefore, after the second translation the point M has been moved to the point M_2'. It is clear that

$$\overrightarrow{MM_2'} = \overrightarrow{MM_1} + \overrightarrow{M_1M_2'}, \tag{3.5}$$

hence

$$\overrightarrow{MM_2'} = \overrightarrow{OA_1} + \overrightarrow{OA_2} = \overrightarrow{OA}, \tag{3.6}$$

since

$$\overrightarrow{MM_1} = \overrightarrow{OA_1} \quad \text{and} \quad \overrightarrow{M_1M_2'} = \overrightarrow{OA_2}. \tag{3.7}$$

Therefore, instead of the two translations we can perform only the one translation defined by the vector $\overrightarrow{MM_2'} = \overrightarrow{OA}$, which is the sum of $\overrightarrow{OA_1}$ and $\overrightarrow{OA_2}$. □

Remarks It is easy to show the following:

(i) A translation of a shape can be replaced by two other consecutive translations.
(ii) The above holds for more than two consecutive translations as well.
(iii) The resultant of the two translations is the translation that replaces them.
(iv) The final position of a shape, when it is a result of several translations, is independent of the order in which the translations take place.

3.2.1 Examples on Translation

Example 3.2.1 Let $ABCD$ be a quadrilateral with $AD = BC$ and let M, N be the midpoints of AB and CD, respectively. Show that MN is parallel to the bisector of the straight semilines AD, BC.

Proof We translate the sides AD and BC to the positions MM_1 and MM_2, respectively (see Fig. 3.8). It is sufficient to show that

$$\widehat{M_1MN} = \widehat{NMM_2}. \tag{3.8}$$

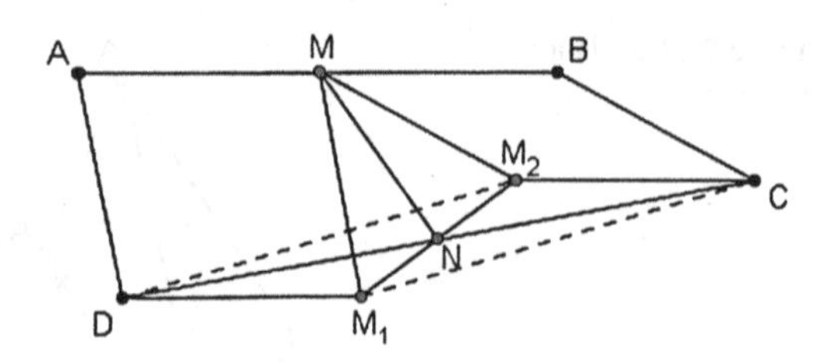

Fig. 3.8 Picture of Example 3.2.1

Indeed, we have

$$MM_1 = MM_2. \tag{3.9}$$

Therefore, MM_1M_2 is an isosceles triangle. Also,

$$\overrightarrow{DM_1} = \overrightarrow{AM} \tag{3.10}$$

because $\overrightarrow{DM_1}$ is a translation of $\overrightarrow{AM}$, and

$$\overrightarrow{M_2C} = \overrightarrow{MB} \tag{3.11}$$

since $\overrightarrow{M_2C}$ is a translation of $\overrightarrow{MB}$.

Additionally, we have

$$\overrightarrow{AM} = \overrightarrow{MB}. \tag{3.12}$$

From Eqs. (3.10), (3.11), and (3.12), we conclude that

$$\overrightarrow{DM_1} = \overrightarrow{M_2C}. \tag{3.13}$$

Therefore, DM_1CM_2 is a parallelogram and thus

$$M_1N = NM_2. \tag{3.14}$$

Therefore, MN is the median of the isosceles triangle MM_1M_2. This means that MN is also the bisector of the angle $\widehat{M_1MM_2}$. □

Example 3.2.2 Let (K, R_1) and (L, R_2) be two circles, ϵ be a straight line and $s > 0$. Are there points A, B on (K, R_1) and (L, R_2), respectively, such that $AB = s$ and $AB \parallel \epsilon$?

Solution Since $AB = s$ and $AB \parallel \epsilon$, we have that B is on the translation of (K, R_1) onto (H, R_1) with $\overrightarrow{KH} = \overrightarrow{AB}$. If (H, R_1) intersects (L, R_2) at the point B, then the vector $\overrightarrow{HB}$ is determined (see Fig. 3.9). Since $\overrightarrow{KH} = \overrightarrow{AB}$, then AB determines the points A and B so that $AB = s$ and $AB \parallel \epsilon$. □

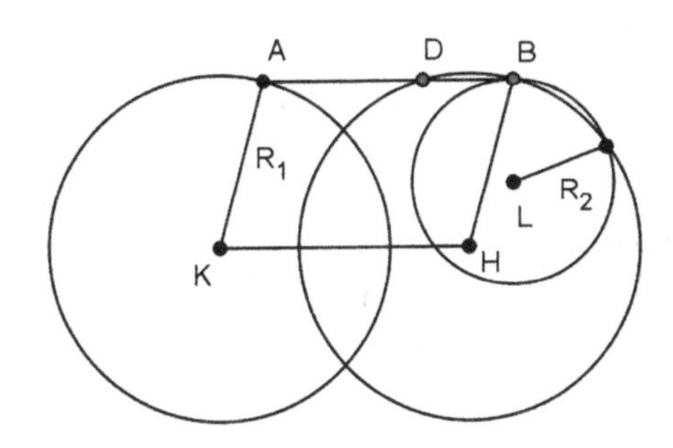

Fig. 3.9 Picture of Example 3.2.2

3.3 Symmetry

3.3.1 Symmetry with Respect to a Center

Let O and M be two points. The point M' is called *symmetrical to the point M with respect to O*, if O is the midpoint of the segment MM'.

Let O be a point and S be a shape. We say that the shape S' is the symmetrical of S with respect to the center O if for every point M of S, there is a point M' of S' such that O is the midpoint of MM', and conversely, if for every point M' of S' there is a point M on S such that O is the midpoint of MM'. Two shapes that are symmetrical to each other are equal.

3.3.2 Symmetry with Respect to an Axis

Let l be a straight line and M be a point. We say that the point M' is *symmetrical to M with respect to the straight line l* if the straight line l is the perpendicular bisector of MM'.

Let l be a straight line and S be a shape. The shape S' is symmetrical to S with respect to l if for every point M of S there is a point M' of S' that is symmetrical to M with respect to l, and conversely, if for every point M' of S' there is a point M on S such that M is the symmetrical point of M' with respect to l.

Two shapes that are symmetrical about an axis are equal.

In the case when the symmetric of each point M, with respect to an axis l, of a shape S lies on S as well, we say that the straight line l is an axis of symmetry of S.

3.4 Rotation

1. An angle $\widehat{xOy}$ is a planar shape. We consider that it is covered by the planar motion of the one side towards the other, while the point O remains fixed. This

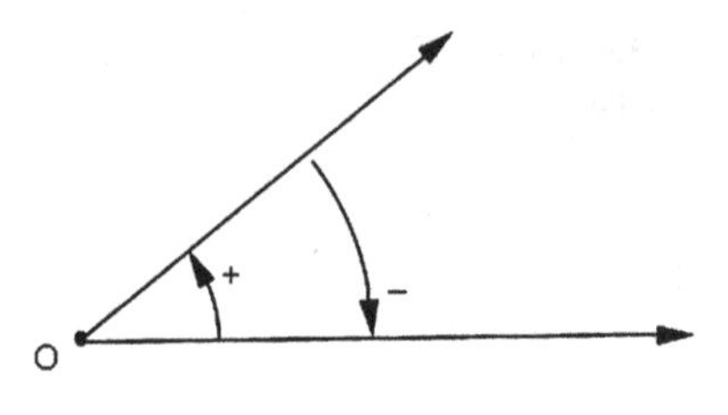

Fig. 3.10 Rotation (Sect. 3.4)

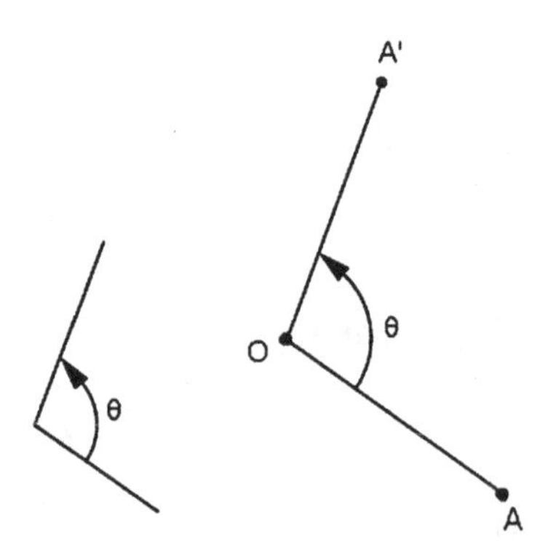

Fig. 3.11 Rotation (Sect. 3.4)

defines an orientation (automatically the opposite orientation can be defined), and in this way we have the sense of a directed angle (see Fig. 3.10).

In particular, consider the plane of the angle $\widehat{xOy}$ and a plane parallel to it on which the arrows of a clock lie. We consider the planar motion of Ox starting at Ox and ending at Oy. If this motion is opposite to the motion of the arrows of the clock, then the angle is considered to be positively oriented. In the opposite case, the angle is considered to be negatively oriented.

2. Let p be a plane and O be a point of p which will be considered as the center of the rotation. Let A be a point of the plane and θ be an oriented angle (see Fig. 3.11). We consider the point A' with

$$\widehat{AOA'} = \theta \tag{3.15}$$

and

$$OA = OA'. \tag{3.16}$$

The correspondence of A to A' is called the *rotation* of A with center O and angle θ. The points A and A' are called *homologous*.

3. The rotation of the shape S with center O and angle θ is the set of the rotated points of S with center O and angle θ.
4. Two shapes such that one is obtained from the other by a rotation about a point O are equal.

Proof Let the shape S' be obtained from the shape S by a rotation about the point O and by an angle θ. We will show that $S = S'$.

Fig. 3.12 Rotation (Sect. 3.4)

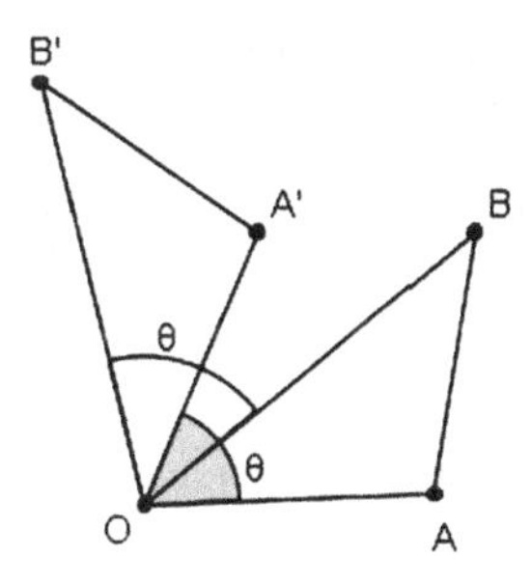

Let A, B be points of S and A', B' be their homologous points (see Fig. 3.12). We have

$$\widehat{AOA'} = \widehat{BOB'} \tag{3.17}$$

and

$$\widehat{AOA'} = \widehat{AOB} + \widehat{BOA'}. \tag{3.18}$$

Also,

$$\widehat{BOB'} = \widehat{BOA'} + \widehat{A'OB'}. \tag{3.19}$$

Therefore,

$$\widehat{AOB} = \widehat{A'OB'}, \tag{3.20}$$

and hence the triangles AOB and $A'OB'$ are equal. Thus

$$AB = A'B', \tag{3.21}$$

for every pair of points A and B. Therefore, $S = S'$. □

5. The rotation of a straight line about the center O and by an angle θ is a straight line. The angle between the two straight lines is θ.
   To show this, we consider two points A and B of the straight line ϵ and their homologous points A' and B', respectively (see Fig. 3.13). We have
$$\widehat{AOA'} = \widehat{BOB'} = \theta. \tag{3.22}$$
   The defined straight line ϵ' is the rotation of ϵ about the center O and by an angle θ.
6. The rotation of a circumference (K, R) about the center O and by an angle θ is the circumference (K', R) equal to (K, R), where K' is the rotation of K. *Hint*. If A is a point on (K, R) and A' its homologous point on (K', R), then the triangles OKA and $OK'A'$ are equal (see Fig. 3.14). From this we conclude that $R = R'$.
7. Generally, let the shape S' be obtained from the shape S by a rotation about the point O by an angle θ. Let A and B be two points of S and let A' and B' be their

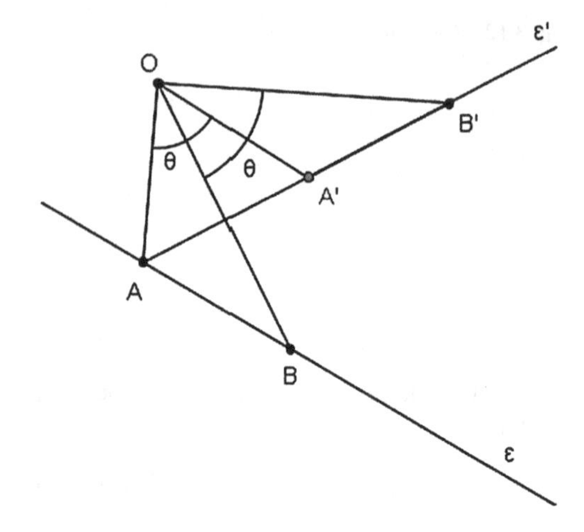

Fig. 3.13 Rotation (Sect. 3.4)

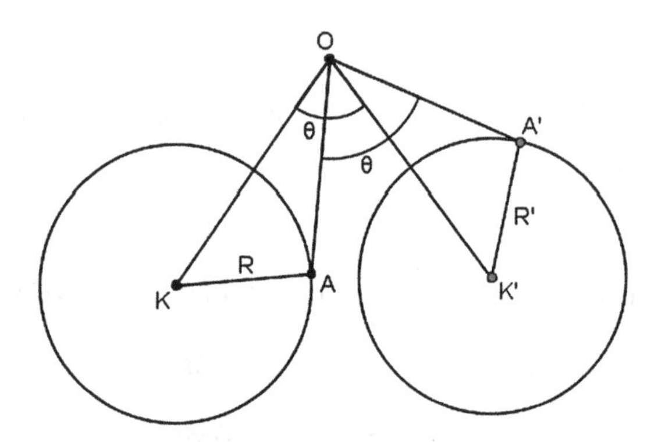

Fig. 3.14 Rotation (Sect. 3.4)

corresponding homologous points which lie on S'. Let l be the straight line that connects A and B and let l' be the straight line that connects the points A' and B'. Then the angle between the straight lines l and l' is equal to the angle θ.
Hint. The proof is based on the previous proposition.

8. If two shapes S, S_1 are equal, then one can be applied onto another by means of a translation and a rotation.

   Indeed, let A, A' be corresponding homologous points of the two equal shapes S, S_1. We translate the first shape S by $\overrightarrow{AA'}$ and bring the point A to the point A'. This translates S to S'. By rotating S' about A', we observe that S' coincides with S_1.

Remark Symmetry with respect to a point O is the same as the rotation about the point O by an angle π.

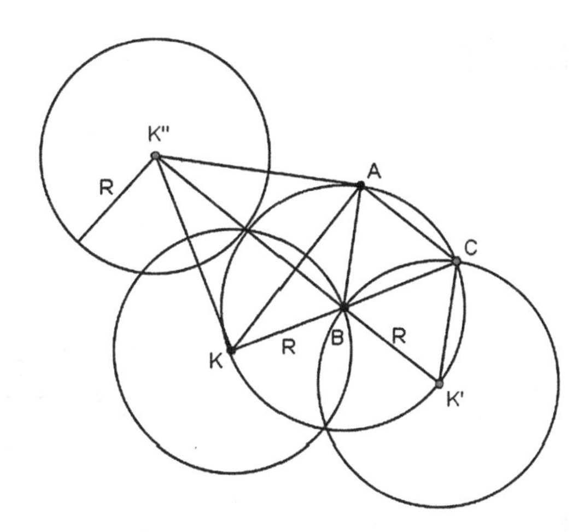

Fig. 3.15 Picture of Example 3.4.1

3.4.1 Examples of Rotation

Example 3.4.1 Let (K,R) be a circle and let A be a point outside the circle. Let B be a point that moves on the circle and let the triangle ABC be moving so that $\widehat{A}=\widehat{B}=\widehat{C}=\frac{\pi}{3}$. Where does the point C lie?

Solution Since

$$AC=AB \tag{3.23}$$

and

$$\widehat{BAC}=\frac{\pi}{3} \quad \text{or} \quad \widehat{BAC}=-\frac{\pi}{3}, \tag{3.24}$$

the point C belongs to the rotation of (K,R) with center A and angle $\pi/3$ or $-\pi/3$, respectively (see Fig. 3.15). □

Example 3.4.2 Consider a shape that moves on a constant plane such that it remains unchanged and each one of two given lines of it pass through a constant point. Show that there are infinitely many lines of the plane, each of which rotates about a constant point.

Proof Let ϵ_1, ϵ_2 be lines passing through the points O_1, O_2, respectively (see Fig. 3.16). Since the shape remains unchanged, the angle $\widehat{(\epsilon_l,\epsilon_2)}=\widehat{\omega}$ remains constant. Therefore, the intersection point B moves on a constant arc whose points see the line segment O_1O_2 under angle $\widehat{\omega}$. A line ϵ passing through B with $\widehat{(\epsilon,\epsilon_1)}=\widehat{\phi}$ intersects the circumference at the point K. Since the angle $\widehat{\phi}$ remains constant, this shows that the line ϵ passes through K. Therefore, each line that passes through B and preserves a constant angle with ϵ_1, passes through a constant point. □

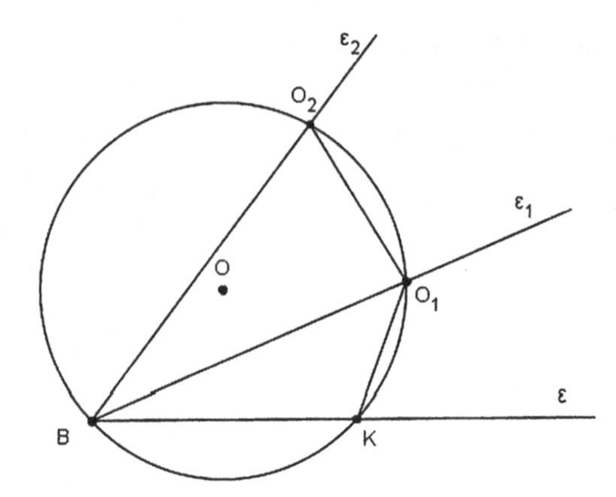

Fig. 3.16 Picture of Example 3.4.2

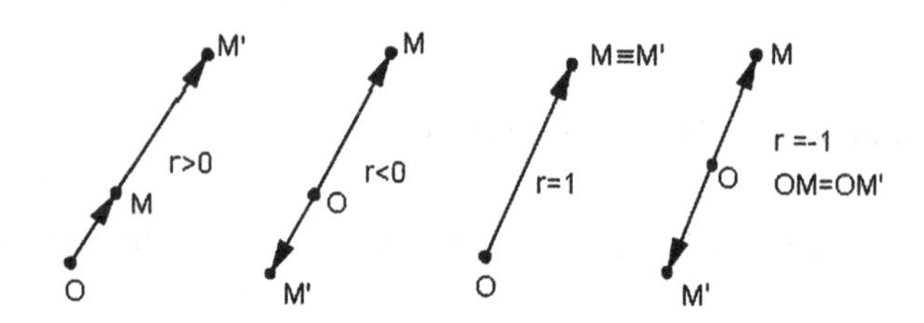

Fig. 3.17 Homothety (Sect. 3.5)

3.5 Homothety

1. The point M' is called *homothetic* to the point M with respect to a point O (the *homothetic center*) if the vector $\overrightarrow{OM'}$ satisfies

$$\overrightarrow{OM'} = r\overrightarrow{OM}, \tag{3.25}$$

where $r \neq 0$. The real number r is called the *ratio of homothety.*

We observe that if $r > 0$, then $\overrightarrow{OM}$ and $\overrightarrow{OM'}$ have the same orientation, and if $r < 0$, the vectors $\overrightarrow{OM}$ and $\overrightarrow{OM'}$ have opposite orientations. If $r > 0$, the point M' is said to be *directly homothetic* to the point M with respect to the point O, and if $r < 0$, then the point M' is said to be *homothetic by inversion* to the point M (see Fig. 3.17).

   (i) The center of homothety is homothetic with respect to itself.
   (ii) The points M, M' are called *homologous* or *corresponding* points.

2. The planar shape S' is homothetic to the planar shape S if there is a real number $r \neq 0$ such that the points of S' are homothetic to the points of S with ratio of homothety r (i.e., S' is the geometrical locus of the homothetic points of S) (see Figs. 3.18, 3.19).

   (i) If $r = 1$ then $S \equiv S'$.
   (ii) If M is a point homothetic to M' with respect to center O and ratio r, then M' is homothetic to M with respect to the point O and ratio $1/r$.

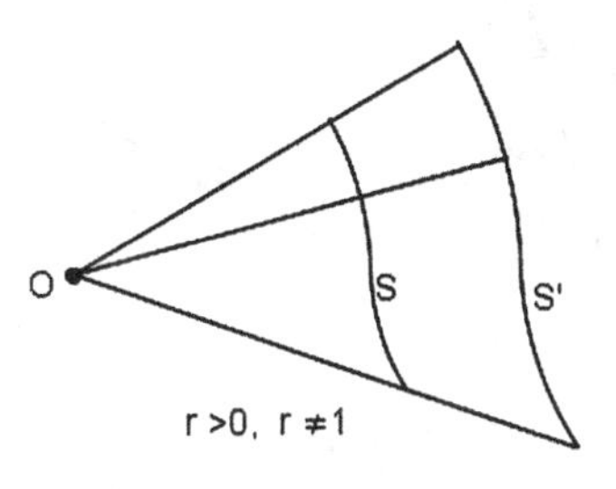

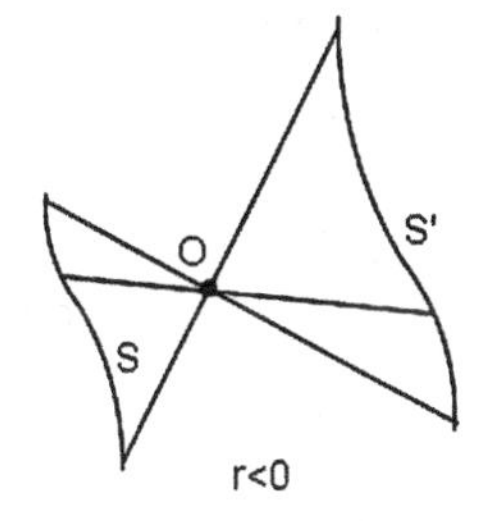

Fig. 3.18 Homothety (Sect. 3.5)

Fig. 3.19 Homothety (Sect. 3.5)

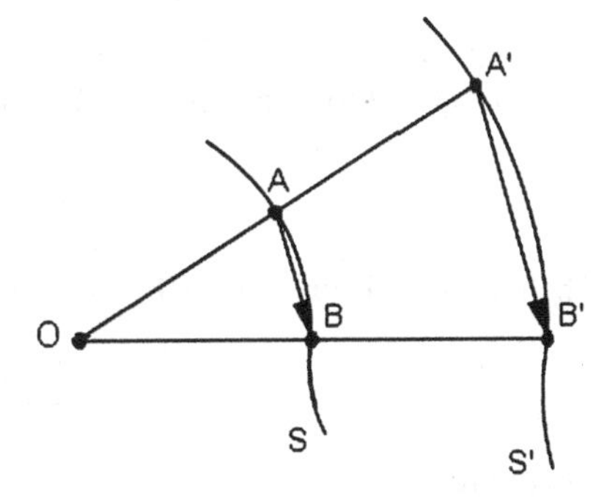

3. *A characteristic criterion of homothety*. A necessary and sufficient condition for a shape S' to be homothetic to a shape S with ratio $r \neq 0, 1$ is that for each pair of points A, B of S there is a pair of points A', B' of S' such that $\overrightarrow{A'B'} = r\overrightarrow{AB}$.

Proof Let S' be homothetic to S with respect to O with ratio $r \neq 0, 1$. Let A, B be points of S and let A', B' be their corresponding homologous points on S'. Then

$$\overrightarrow{OA'} = r\overrightarrow{OA} \tag{3.26}$$

and

$$\overrightarrow{OB'} = r\overrightarrow{OB}. \tag{3.27}$$

Therefore,

$$\overrightarrow{OA'} - \overrightarrow{OB'} = r(\overrightarrow{OA} - \overrightarrow{OB}), \tag{3.28}$$

that is,

$$\overrightarrow{B'A'} = r\overrightarrow{BA}, \tag{3.29}$$

or equivalently,

$$\overrightarrow{A'B'} = r\overrightarrow{AB}. \tag{3.30}$$

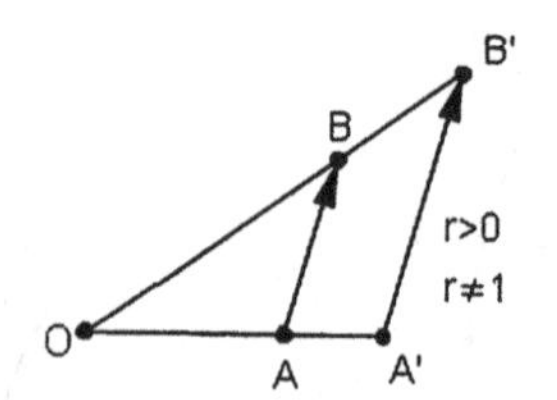

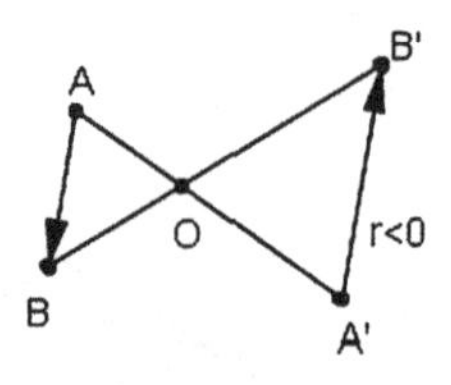

Fig. 3.20 Homothety (Sect. 3.5)

Conversely, let A, B be points on S and let A', B' be points on S' such that

$$\overrightarrow{A'B'} = r\overrightarrow{AB}, \tag{3.31}$$

with $r \neq 0, 1$. Let A, A' be constant points and B, B' traverse S, S', respectively. We consider O to be a point of AA' so that

$$\overrightarrow{OA'} = r\overrightarrow{OA}. \tag{3.32}$$

The number r is unique. By hypothesis, we have

$$\overrightarrow{A'B'} = r\overrightarrow{AB}. \tag{3.33}$$

Therefore,

$$\overrightarrow{OA'} + \overrightarrow{A'B'} = r(\overrightarrow{OA} + \overrightarrow{AB}), \tag{3.34}$$

that is,

$$\overrightarrow{OB'} = r\overrightarrow{OB}, \tag{3.35}$$

for the points B, B' with $\overrightarrow{A'B'} = r\overrightarrow{AB}$. □

Corollary 3.1 *If the point M "produces" the vector $\overrightarrow{AB}$, then the point M' which is homothetic to M produces the vector $\overrightarrow{A'B'}$ which is homothetic to $\overrightarrow{AB}$.*

Corollary 3.2 *The homothetic shape of a line is a line parallel to the initial one (see Fig.* 3.20).

Corollary 3.3 *The homothetic shape of a planar polygon is a polygon similar to the initial one. Its sides have the same orientation with the sides of the initial polygon if $r > 0$ and the opposite orientation if $r < 0$.*

Remark 3.1 The converse of Corollary 3.3 is also true, that is, if two similar planar polygons have respective sides that all have the same or all have the opposite orientation, then the polygons are homothetic. If $r \neq 0$, then there is a center of homothety.

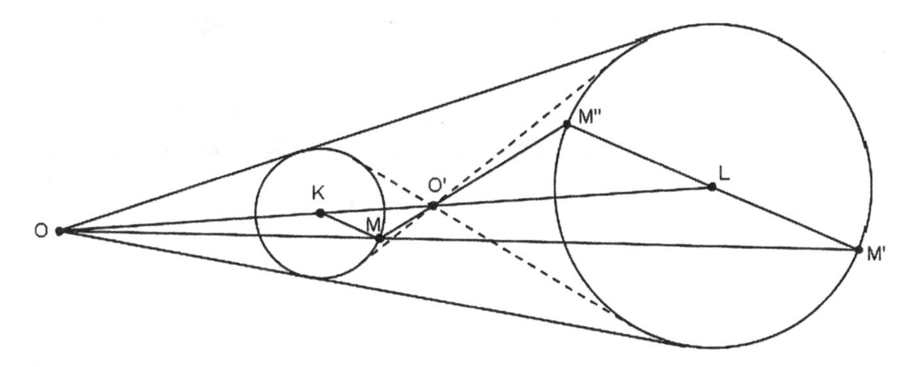

Fig. 3.21 Homothety (Sect. 3.5)

Remark Suppose that the points M_i form a planar polygon and that the points S_i also form a planar polygon ($i = 3, 4, \ldots, n$) where n is a natural number. Then if it is asked whether the lines $M_i S_i$ have a common point, then that point is very likely going to be the center of homothety.

Theorem 3.2 *If two planar polygons are similar, then they can be positioned so that they are homothetic.*

Hint. Let S, S' be two planar similar polygons. Let A, B be two vertices of the polygon S and A', B' be the corresponding vertices of the polygon S'. If we consider a point K such that $A'K \parallel AB$, it suffices to rotate the polygon S' about the point A' by the angle $\widehat{B'A'K}$.

4. Two circles are homothetic shapes with ratio r equal to the ratio of their radii.
   (i) The centers of the circles are homologous points.
   (ii) The circles are homothetic in only two ways if $r \neq 1$. The first homothety has center O and the second one has center O'.

Corollary 3.4 *The common external tangents of two circles (if they exist) pass through the "external" center of homothety which is the intersection of the lines* KL *and* MM' *with* $\overrightarrow{KM}$ *and* $\overrightarrow{LM'}$ *being parallel and having the same orientation. The common internal tangents of two circles pass through the "internal" center of homothety which is the intersection of the lines* KL *and* MM'' *with the vectors* $\overrightarrow{KM}$ *and* $\overrightarrow{LM''}$ *being parallel and having opposite orientations.*

Remark 3.2 Two circles with different centers and equal radii do not have an external center of homothety, but have an internal center which is equal to the midpoint of the straight line segment determined by the centers of the circles (see Fig. 3.21 and 3.22).

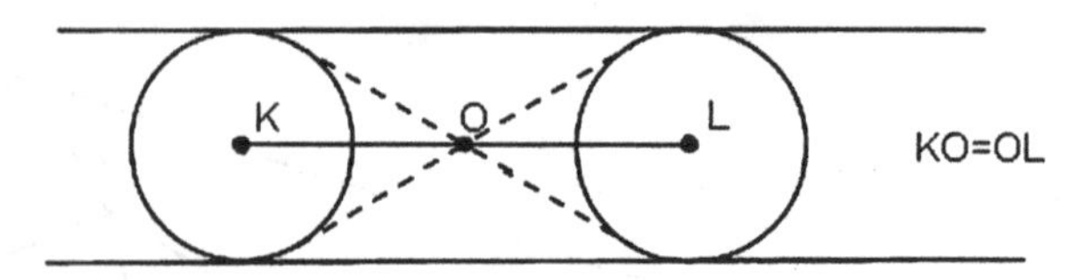

Fig. 3.22 Homothety (Sect. 3.5)

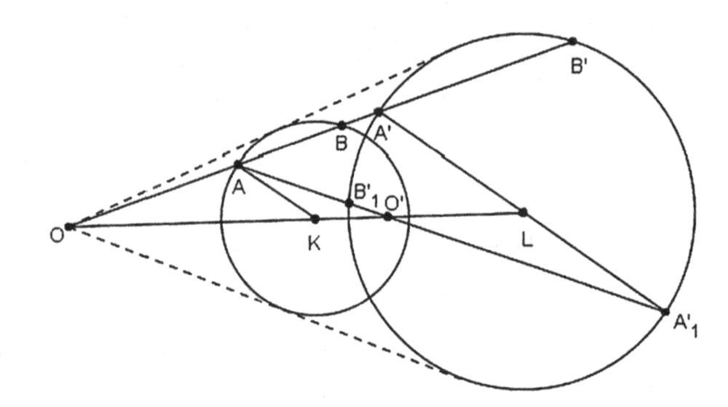

Fig. 3.23 Homothety (Sect. 3.5)

Proposition 3.1 *Let S_1, S_2, S_3 be three shapes. If S_1 is homothetic to S_2 and S_2 is homothetic to S_3, then S_1 is also homothetic to S_3. Furthermore, the three centers of homothety are collinear.*

Hint. Let S_1, S_2, S_3 be three figures such that S_2 is homothetic to S_1 with homothety center O_1 and S_3 is homothetic of S_2 with homothety center O_2. It follows that S_3 shall be homothetic to S_1 with homothety center a certain point O_3. If M_1, M_2 are homological points of the figures S_1, S_2, respectively, and M_3, M_2 are homological points of the figures S_2, S_3, respectively, then we can apply Menelaus' Theorem (4.12) to the triangle $M_1M_2M_3$.

Remark 3.3 The above proposition can be used also as a method for proving that three given points are collinear.

Definition 3.2 Let O be the center of homothety of two circles (K, r_1) and (L, r_2). Let $A \in (K, r_1)$ and $B \in (L, r_2)$ two homologous points of the circles. If the line AB intersects the circles at the points $A_1 \in (K, r_1)$ and $B_1 \in (L, r_2)$, the point B_1 is called *anti-homologous* of A and the point A_1 is called *anti-homologous* of B.

Theorem 3.3 *The inner product of two vectors with initial point the center of homothety of two circles (K, r_1), (L, r_2) and terminal points a pair of anti-homologous points is constant.*

Proof Since the points A and A' are homologous (Fig. 3.23), we have

$$\frac{OA}{OA'} = \frac{r_1}{r_2}, \tag{3.36}$$

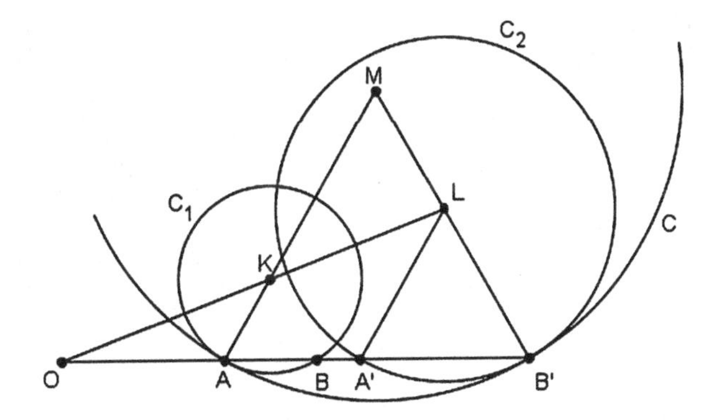

Fig. 3.24 Homothety (Sect. 3.5)

and therefore,

$$OA = \frac{r_1}{r_2} OA'. \tag{3.37}$$

Furthermore,

$$\overrightarrow{OA'} \cdot \overrightarrow{OB'} = OL^2 - r_2^2. \tag{3.38}$$

Thus

$$\overrightarrow{OA} \cdot \overrightarrow{OB'} = \frac{r_1}{r_2}\left(OL^2 - r_2^2\right), \tag{3.39}$$

which is constant. With respect to the center of homothety O', we get

$$\frac{O'A}{O'A_1'} = -\frac{r_1}{r_2} \quad \text{or} \quad \overrightarrow{O'A} = -\frac{r_1}{r_2}\overrightarrow{O'A_1'}. \tag{3.40}$$

Furthermore,

$$\overrightarrow{O'B_1'} \cdot \overrightarrow{O'A_1'} = O'L^2 - R^2. \tag{3.41}$$

Therefore,

$$\overrightarrow{O'A} \cdot \overrightarrow{O'B_1'} = \frac{r_1}{r_2}\left(R^2 - O'L^2\right), \tag{3.42}$$

which is constant. □

Theorem 3.4 *Let C_1, C_2 be two circles and let A, B' be anti-homologous points that belong to C_1, C_2, respectively. Then there exists a circle tangent to C_1 and C_2 at the points A, B', respectively.*

Proof Let O be the homothety center of the circles C_1, C_2 (see Fig. 3.24),

$$M \equiv AK \cap B'L, \qquad B \equiv C_1 \cap AB' \quad \text{and} \quad A' \equiv C_2 \cap AB'.$$

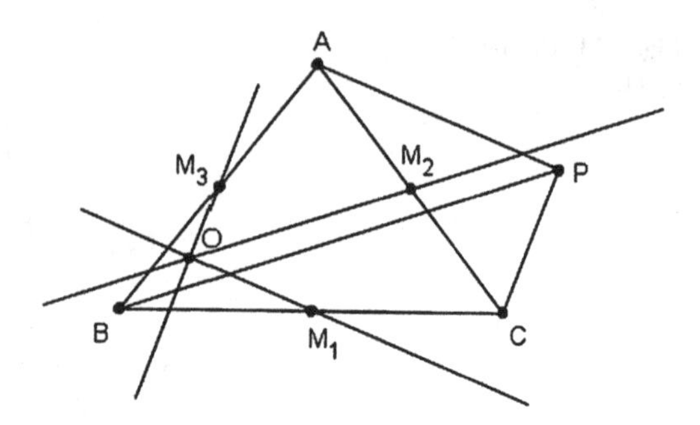

Fig. 3.25 Picture of Example 3.5.1

From the isosceles triangle $LA'B'$, we get

$$\widehat{A'} = \widehat{B'}.$$

But

$$LA' \parallel KA,$$

hence

$$\widehat{A} = \widehat{A'}.$$

Finally, we have

$$\widehat{A} = \widehat{B'}.$$

Thus the triangle MAB' is isosceles. Consequently, there is a circle with center M and radius MA which is tangent to the circles C_1, C_2 at the points A and B'.

We follow the same method with the center of homothety O', which lies between the points K and L. □

3.5.1 Examples of Homothety

Example 3.5.1 Let ABC be a triangle and P be a point on the plane of the triangle. From the midpoint M_1 of BC we draw the line parallel to PA. From the midpoint M_2 of CA we draw the parallel to PB, and from the midpoint M_3 of AB we draw the line parallel to PC. Prove that the three parallel lines pass through the same point.

Proof Let O be the intersection point of the lines parallel to AP, CP that pass through the points M_1 and M_3, respectively (see Fig. 3.25). Then the triangles OM_1M_3 and PAC are homothetic with ratio $r = -2$. Therefore, the center of homothety is the barycenter G of the triangle $M_1M_2M_3$ since, if M_2' is the midpoint of M_1M_3, we have

$$\overrightarrow{GM_2} = -2\overrightarrow{GM_2'}. \tag{3.43}$$

Therefore,

$$\overrightarrow{GO} = -\frac{1}{2}\overrightarrow{GP}. \tag{3.44}$$

Consequently, the point O is the point of intersection of the three lines. □

Example 3.5.2 Let (K_1, r_1), (K_2, r_2), and (K_3, r_3) be circles. Let O_1, O_1' be the centers of homothety of the circles (K_3, r_3), (K_2, r_2), let O_2, O_2' be the centers of homothety of the circles (K_1, r_1), (K_3, r_3), and O_3, O_3' be the centers of homothety of the circles (K_1, r_1), (K_2, r_2). Prove that the lines K_1O_1', K_2O_2', and K_3O_3' pass through the same point.

Proof For the triangle $K_1K_2K_3$, we obtain

$$\frac{O_2'K_1}{O_2'K_3} = \frac{r_1}{r_3}, \tag{3.45}$$

$$\frac{O_3'K_2}{O_3'K_1} = \frac{r_2}{r_1}, \tag{3.46}$$

and

$$\frac{O_1'K_3}{O_1'K_2} = \frac{r_3}{r_2}. \tag{3.47}$$

Therefore,

$$\frac{O_2'K_1}{O_2'K_3} \cdot \frac{O_3'K_2}{O_3'K_1} \cdot \frac{O_1'K_3}{O_1'K_2} = 1. \tag{3.48}$$

Hence, Ceva's theorem applies (see Chap. 4: Theorems), and therefore the lines K_1O_1', K_2O_2', and K_3O_3' pass through the same point. □

Example 3.5.3 Let $ABCD$ be a rectangle such that $AB = BC\sqrt{2}$. Let E be a point of the semicircumference with diameter AB which does not have common part with $ABCD$ apart from AB. Let K, L be the intersections of AB with ED and EC, respectively. Show that

$$AL^2 + BK^2 = AB^2. \tag{3.49}$$

Proof We consider a rectangle $KLNM$, homothetic to the rectangle $ABCD$ (see Fig. 3.26). Then the points A, M, E are collinear, and the points B, N, E are collinear. Let

$$AK = a, \qquad KL = b, \quad \text{and} \quad BL = c.$$

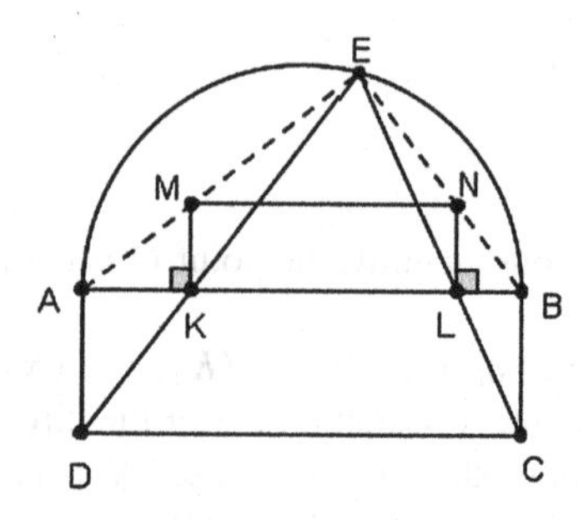

Fig. 3.26 Picture of Example 3.5.3

Equation (3.49) is equivalent to

$$(a+b)^2+(b+c)^2=(a+b+c)^2, \tag{3.50}$$

which is equivalent to

$$b^2=2ac. \tag{3.51}$$

It is therefore enough to show that Eq. (3.51) holds true. Because of the homothety, we have that *ABCD* and *MNLK* are similar. Therefore,

$$KL=MK\sqrt{2}, \tag{3.52}$$

and from the similarity of the triangles *AMK* and *BNL*, we have

$$\frac{a}{MK}=\frac{NL}{c}, \tag{3.53}$$

therefore,

$$MK^2=ac, \tag{3.54}$$

and thus

$$2MK^2=2ac. \tag{3.55}$$

Hence

$$KL^2=2ac, \tag{3.56}$$

and thus

$$b^2=2ac. \tag{3.57}$$

□

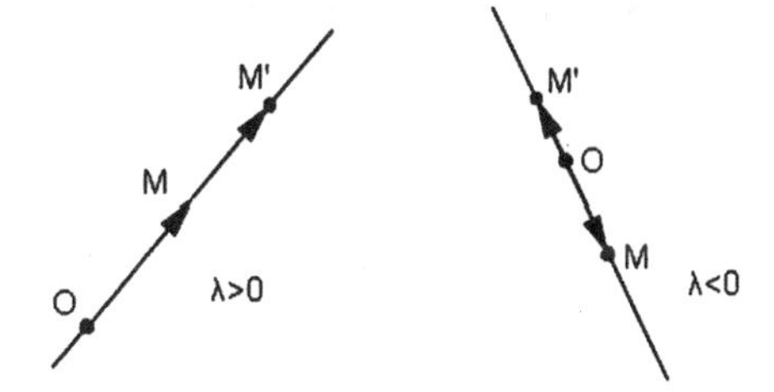

Fig. 3.27 Inversion (Sect. 3.6.1)

3.6 Inversion

3.6.1 Inverse of a Point

Let O be a point in the Euclidean plane E^2, $\lambda \in \mathbb{R} \setminus \{0\}$. The *inverse* of a point M with respect to O is the point $M' \in OM$ such that

$$\overrightarrow{OM} \cdot \overrightarrow{OM'} = \lambda$$

where "·" stands for the usual inner product of two vectors. The point O is said to be the *pole* (or *inversion center*) and the real number λ the *power* of the inversion. If $\lambda > 0$ then $\overrightarrow{OM}$, $\overrightarrow{OM'}$ are of the same *orientation* $(\widehat{(\overrightarrow{OM}, \overrightarrow{OM'})} = 0)$.

In the case $\lambda < 0$, $\overrightarrow{OM}$, $\overrightarrow{OM'}$ are of *opposite* orientation $(\widehat{(\overrightarrow{OM}, \overrightarrow{OM'})} = \pi)$.

The inverse of a point M with respect to a pole O and of power $\lambda \in \mathbb{R} \setminus \{0\}$ is uniquely defined (see Fig. 3.27).

The inverse of the pole O with respect to itself and of power $\lambda \in \mathbb{R} \setminus \{0\}$ is a point at *infinity*.

3.6.2 Inverse of a Figure

Let a figure S, a point O, and a real number $\lambda \neq 0$ be given. Define the *inverse of* S with respect to the pole O and of power λ to be a figure S' that is the locus of the inverses of the points of the figure S with respect to the pole O and of the same power λ.

It is evident that the property of inversibility satisfies the duality condition: *If* S' *is the inverse of* S *with respect to the pole* O *and of power* λ, then S *is the inverse of* S' *with respect to the same pole* O *and of the same power*. In order to abbreviate notation, we shall denote, in what follows, by $\text{Inv}_{(O,\lambda)}\, S = S'$ *the inverse of* S *with respect to the pole* O *and of power* λ.

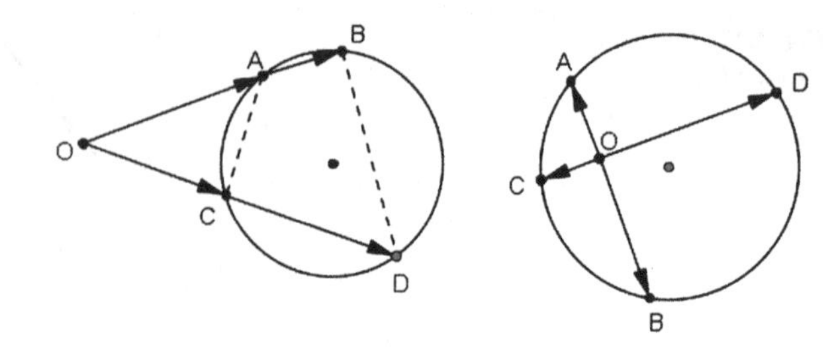

Fig. 3.28 Criterion (Sect. 3.6.4)

3.6.3 An Invariance Property

Let ϵ be a straight line, O a point, and $\lambda \neq 0$. Then, $\mathrm{Inv}_{(O,\lambda)}\,\epsilon = \epsilon$ if and only if $O \in \epsilon$. Indeed, if $M \in \epsilon$ and M' is its inverse with respect to O of power λ, then the points M, O, M' are collinear with $O, M \in \epsilon$ and thus $M' \in \epsilon$.

3.6.4 Basic Criterion

It is well known that the power of a point O with respect to a circle $C(K, \rho)$ is the product $\overrightarrow{OA} \cdot \overrightarrow{OB}$, with O, A, B collinear and A, B points of the circle. Furthermore, if C and D are points of the circle with O, C, D collinear and belonging to a straight line, different from the one defined by A and B, then

$$\overrightarrow{OA} \cdot \overrightarrow{OB} = \overrightarrow{OC} \cdot \overrightarrow{OD}.$$

Conversely, let $O \in E^2$, ϵ_1, ϵ_2 ($\epsilon_1 \neq \epsilon_2$) be straight lines with $\{O\} = \epsilon_1 \cap \epsilon_2$ and $A, B \in \epsilon_1$, $C, D \in \epsilon_2$ such that

$$\overrightarrow{OA} \cdot \overrightarrow{OB} = \overrightarrow{OC} \cdot \overrightarrow{OD},$$

then the points A, B, C, D belong to the same circle (see Fig. 3.28). Thus we derive the following

Corollary 3.5 *Let S, S' be two figures in the Euclidean plane E^2, $S' = \mathrm{Inv}_{(O,\lambda)}\,S$, $\lambda \neq 0$, and O the pole of inversion. Then, for any pair of points $A, C \in S$ with corresponding inverses $B, D \in S'$ the points A, B, C, D belong to the same circle.*

Conversely, let O be a point and ϵ_1, ϵ_2 ($\epsilon_1 \neq \epsilon_2$) be two straight lines. Let S, S' be a pair of plane figures and $O \notin S \cap S'$ be a given point. Suppose that for any pair of points $A, B \in S$ the corresponding inverse images $C, D \in S'$ are obtained by means of the corresponding intersections of the straight lines OA, OC with S' so that O, A, B, C and D are homocyclic, then the figure S' is the inverse of S with respect to the pole O and power $\overrightarrow{OA} \cdot \overrightarrow{OB}$.

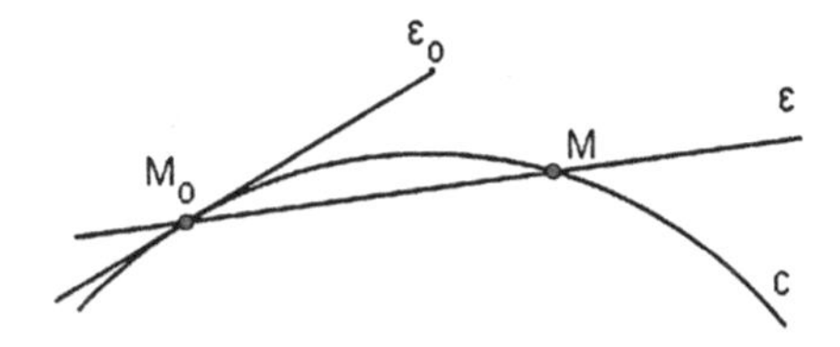

Fig. 3.29 Tangent to a curve (Sect. 3.6.7)

3.6.5 Another Invariance Property

Let C be a circle and O a point in the plane of the circle. The point O is considered to be the pole of the inversion with power a real number $\lambda \neq 0$. Then, the circle C admits as its inverse C itself if and only if the power of the pole O with respect to the circle is equal to the power λ of the inversion.

3.6.6 Invertibility and Homothety

Theorem 3.5 *Two figures S_1 and S_2 which are the inverses of a third figure $\tilde{S}$ with respect to the same pole of inversion O are homothetic.*

Proof Let S_1, S_2 be two figures with $S_i = \mathrm{Inv}_{(O,\lambda_i)}\, \tilde{S}$, and λ_i be the corresponding powers of inversion where $\lambda_i \in \mathbb{R} \setminus \{0\}$ for $i = 1, 2$. Consider the points $M_i \in S_i$, $i = 1, 2$ with $M_i = \mathrm{Inv}_{(O,\lambda_i)}\, M$ for a certain point $M \in \tilde{S}$. By the above assumptions, the following properties hold:

$$\begin{array}{ccc} \overrightarrow{OM_1} \cdot \overrightarrow{OM} = \lambda_1 & & OM_1 \cdot OM = |\lambda_1|, \\ & \Rightarrow & \\ \overrightarrow{OM_2} \cdot \overrightarrow{OM} = \lambda_2 & & OM_2 \cdot OM = |\lambda_2|. \end{array} \tag{3.58}$$

Hence, we obtain

$$\frac{OM_1}{OM_2} = \frac{|\lambda_1|}{|\lambda_2|}. \tag{3.59}$$

Therefore, the figures are homothetic. □

3.6.7 Tangent to a Curve and Inversion

Let $c : I \to E^2$, $I \subset \mathbb{R}$, be a plane curve and the point $M_0 \in c$. For any point $M \in c$, an infinity of straight lines M_0M can be constructed and, intuitively speaking, the limit, if it exists, of this family of straight lines passing through the point M_0 shall be a straight line ϵ_0 which is called the *tangent* of the curve c at M_0 and it has the property (see Fig. 3.29): There exists a segment AB of c, with M_0 being an

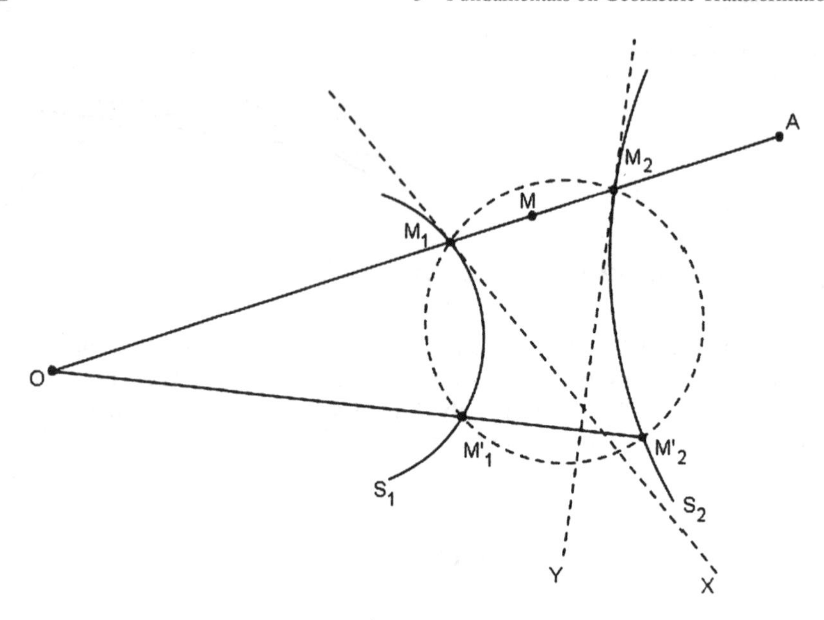

Fig. 3.30 Tangent to a curve (Sect. 3.6.7)

interior point of AB, such that M_0 is the unique common point of ϵ_0 and AB. The behavior of the tangents of a curve with respect to the operation of inversion can be characterized by the following:

Theorem 3.6 *Let S_1, S_2 be a pair of plane curves such that $S_1 = \mathrm{Inv}_{(O,\lambda)}\, S_2$, $\lambda \neq 0$. The tangent lines to the corresponding points M_i of S_i, $i = 1, 2$ form angles with $\overrightarrow{M_1 M_2}$ that are equal.*

Proof We have $S_1 = \mathrm{Inv}_{(O,\lambda)}\, S_2$, $\lambda \neq 0$. It should be enough to prove that $\widehat{XM_1M_2} = \widehat{M_1M_2Y}$ where M_1X is tangent to S_1 at the point M_1 and M_2Y is tangent to S_2 at the point M_2, and the points O, M_1, M_2, M, A are collinear. For any point M_1' of the curve S_1, there exists the corresponding (inverse) point $M_2' \in S_2$. Using Theorem 3.5, the quadrilateral $M_1M_1'M_2'M_2$ can be inscribed in a circle and thus it holds (see Fig. 3.30)

$$\widehat{M_1M_1'M_2'} = \widehat{M_2'M_2A}.$$

Even to the limit, this property of inscribability still holds, hence

$$M_1 \to M_1' \quad \Leftrightarrow \quad M_2 \to M_2'.$$

In this case, we deduce that

$$\widehat{XM_1M_2} = \widehat{M_1M_2Y}.$$

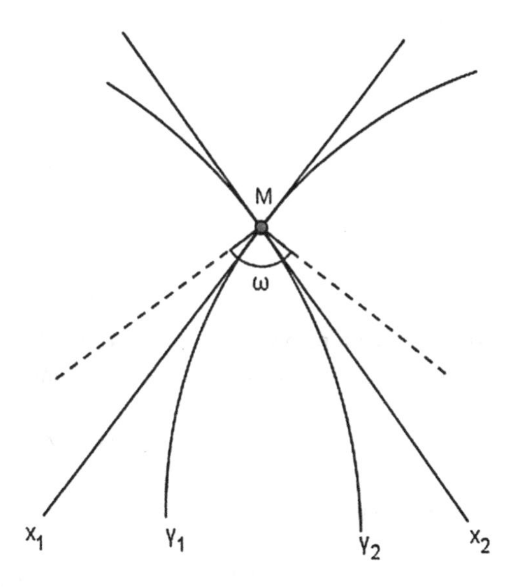

Fig. 3.31 Tangent to a curve (Sect. 3.6.8)

Therefore, if the tangents are intersecting at a point P then the triangle PM_1M_2 has to be isosceles. In general, these tangents have to be symmetrical with respect to the axis of symmetry determined by the perpendicular bisector of the line segment M_1M_2. □

3.6.8 Inversion and Angle of Two Curves

Let y_1, y_2 be two curves intersecting at a point M. We define the angle $\widehat{y_1My_2}$ (of the curves at their common point M) to be the complement of the angle formed by their semi-tangents at this point M. The following holds:

Theorem 3.7 *The angle of two intersecting curves at a point M is equal (in measure) to either the angle formed by the intersection of the corresponding inverse curves at the corresponding point M' or to their symmetric counterparts with respect to the perpendicular line at the middle point of MM'.*

Proof It is an immediate consequence of Theorem 3.6 (see Fig. 3.31). □

3.6.9 Computing Distance of Points Inverse to a Third One

Let the point O of the Euclidean plane E^2 be the inversion pole, $\lambda \in \mathbb{R} \setminus \{0\}$ the inversion power, and S_1, S_2 be two figures which are inverse to each other with respect to the point O and of power λ. Let also $A, B \in S_1$ be given and their corresponding

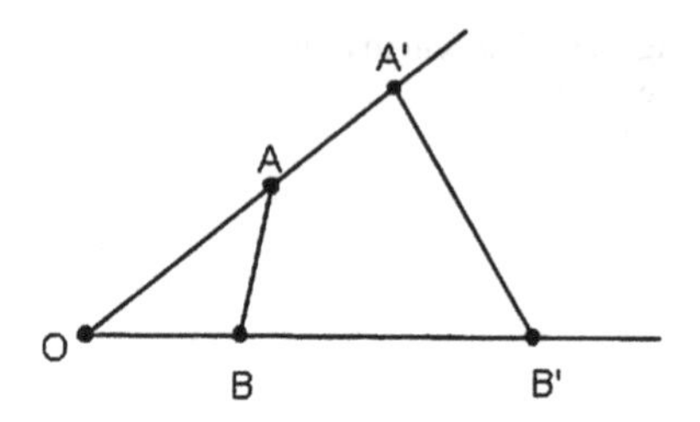

Fig. 3.32 Distance (Sect. 3.6.9)

λ-*inverses* with respect to the point O be $A', B' \in S_2$. It is evident that the quadrilateral $AA'B'B$ can be inscribed in a circumference and, in fact (see Fig. 3.32),

$$\triangle OAB \sim \triangle OA'B'.$$

It follows that

$$\frac{A'B'}{AB} = \frac{OA'}{OB} \quad \Rightarrow \quad A'B' = AB \cdot \frac{OA'}{OB}. \tag{3.60}$$

But

$$\overrightarrow{OA'} \cdot \overrightarrow{OA} = \lambda \quad \Rightarrow \quad OA' \cdot OA = |\lambda|. \tag{3.61}$$

From (3.60) and (3.61), we derive that the distance between the inverses A', B' is given by

$$A'B' = AB \cdot \frac{OA' \cdot OA}{OB \cdot OA} = AB \cdot \frac{|\lambda|}{OA \cdot OB}. \tag{3.62}$$

3.6.10 Inverse of a Line Not Passing Through a Pole

Let ϵ be a straight line in the Euclidean plane E^2, then the inverse of ϵ with respect to the pole O with $O \notin \epsilon$ is a circle C_ϵ passing through the pole O. The diameter of the circle passing through O is perpendicular to ϵ.

In fact, this holds true because

$$\overrightarrow{OM} \cdot \overrightarrow{ON} = \overrightarrow{OT} \cdot \overrightarrow{OP}, \tag{3.63}$$

and therefore, either *MTNP* is a quadrilateral with $\widehat{N} = \widehat{T} = 90°$ or *MTNP* is inscribed in a certain circle, that is, $\widehat{ONP} = 90°$, and thus *the point N is moving on a circle of diameter OP* (see Figs. 3.33 and 3.34). Consequently, $\epsilon \perp OP$.

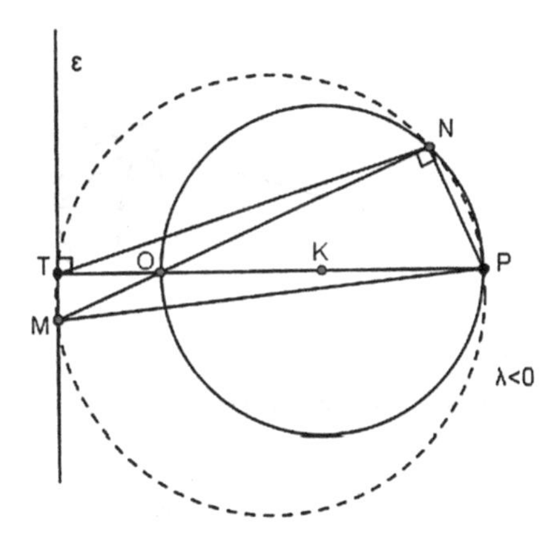

Fig. 3.33 Inverse of a line (Sect. 3.6.10)

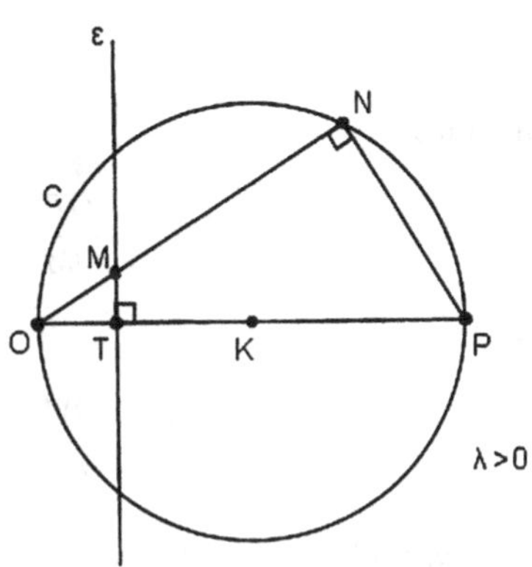

Fig. 3.34 Inverse of a line (Sect. 3.6.10)

3.6.11 Inverse of a Circle with Respect to a Pole Not Belonging to the Circle

Let $C(K, \rho)$ be a circle in the Euclidean plane E^2, and O a point with $O \notin C(K, \rho)$. We consider the point O as the inversion pole with power $\lambda \neq 0$. We are going to determine the curve described by the inverse M' of the point M when M runs along the circle $C(K, \rho)$. Let OK be the straight line intersecting the circle C at the points T and Y. It is true that

$$\overrightarrow{OM} \cdot \overrightarrow{OM'} = \lambda. \tag{3.64}$$

Consider the points $T', Y' \in OK$ such that

$$\overrightarrow{OT} \cdot \overrightarrow{OT'} = \lambda \tag{3.65}$$

and

$$\overrightarrow{OY} \cdot \overrightarrow{OY'} = \lambda. \tag{3.66}$$

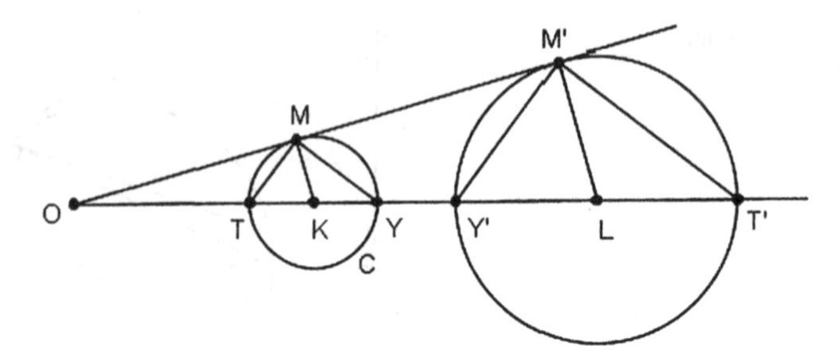

Fig. 3.35 Inverse of a circle (Sect. 3.6.11)

Using (3.64), (3.65), and (3.66), we derive (see Fig. 3.35)

$$OM \cdot OM' = OT \cdot OT' = OY \cdot OY' = |\lambda|, \tag{3.67}$$

and thus

$$\frac{OM}{OT} = \frac{OT'}{OM'} \quad \Rightarrow \quad \Delta OMT \sim \Delta OM'T' \tag{3.68}$$

and

$$\frac{OM}{OY} = \frac{OY'}{OM'} \quad \Rightarrow \quad \Delta OMY \sim \Delta OM'Y'. \tag{3.69}$$

However, $\widehat{YMT} = 90°$, hence $\widehat{T'M'Y'} = 90°$ and the geometrical locus of the point M' has to be a circle of diameter equal to $T'Y'$. This is actually the inverse of the circle C.

3.6.12 Inverse of a Figure Passing Through the Pole of Inversion

The inverse of a circle passing through the pole of inversion is a straight line perpendicular to the diameter of the circle passing through the center—the pole of the inversion (see Fig. 3.36). Indeed, it is enough to observe that the relation

$$OM \cdot OM' = OA \cdot OB \tag{3.70}$$

holds true. This is the case when the point A is the foot of the perpendicular from the point M' to the line OB, since either the quadrilateral $ABMM'$ can be inscribed in a circle or the quadrilateral $AMBM'$ can be inscribed in a circle. This happens because either

$$\widehat{OMB} = \widehat{BAM'} = 90° \quad \text{when } \overrightarrow{OM} \cdot \overrightarrow{OM'} = p > 0,$$

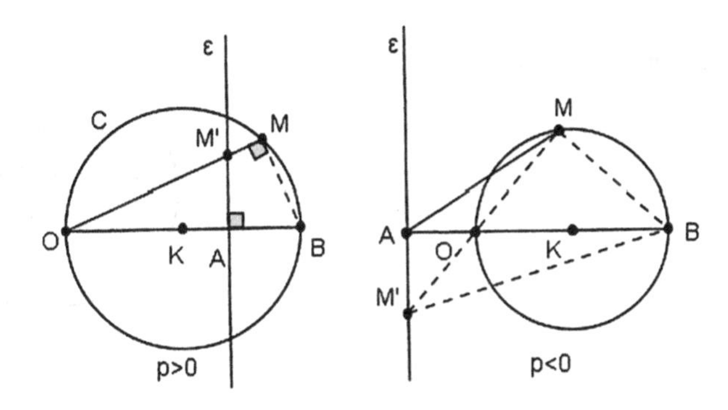

Fig. 3.36 Inverse of a figure (Sect. 3.6.12)

or

$$\widehat{M'AB} = \widehat{M'MB} = 90^\circ \quad \text{when } \overrightarrow{OM} \cdot \overrightarrow{OM'} = p < 0,$$

respectively. The following holds:

Theorem 3.8 *Two circles can always be considered inverses to one another in exactly two different ways if they are not tangent and in exactly one way if they are tangent.*

Proof Indeed, when the circles are not tangent, this occurs since two circles with centers O_1, O_2 are homothetic in exactly two different ways, in general. This means that they can be considered inverses to one another in two different ways. The pole of the inversion is the same with the center of homothety, and the power of inversion is equal to the product of the powers of the poles with the similarity ratio, with respect to the first circumference. □

Remark 3.4 These two inversions are the only operations that transform one of the circles under consideration to the other, and vice versa.

3.6.13 Orthogonal Circles and Inversion

Let an inversion of pole O with power $\rho > 0$ be given. The circle of center O and radius $\sqrt{\rho}$ is the geometrical locus of the points of the Euclidean plane E^2 which coincide with their inverses. This is called the *inversion circle of pole O and of power ρ.*

Theorem 3.9 *Any circle $C_2(K, R')$ passing through a pair of inverse points A, B is orthogonal to the inversion circle $C_1(O, R)$.*

Proof It is enough to observe that (see Fig. 3.37)

$$\overrightarrow{OA} \cdot \overrightarrow{OB} = R^2, \qquad |\overrightarrow{OP}| = R.$$

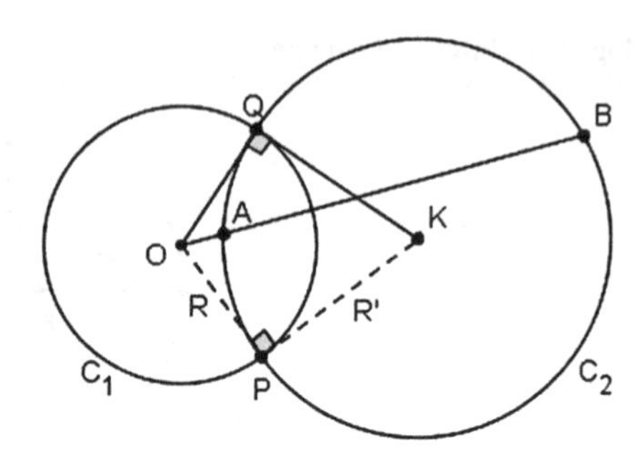

Fig. 3.37 Orthogonal circles and inversion (Sect. 3.6.13)

Thus

$$\overrightarrow{OP}^2 = \overrightarrow{OA} \cdot \overrightarrow{OB},$$

which implies

$$\widehat{OPK} = 90^\circ. \qquad \square$$

Remark 3.5 If the points A, B are such that every circle of center K passing through A and B is orthogonal to every other circle of center O and of radius R, then the points A, B are inverses with respect to this circle.

3.6.14 Applications of the Inversion Operation

Example 3.6.1 (Ptolemy's inequality) Let A, B, C, D be four points in the plane, then

$$AC \cdot BD \leq AB \cdot DC + AD \cdot BC. \tag{3.71}$$

Proof Consider the inversion with pole A and power a certain real number $\rho \neq 0$. Let B', C', and D' be the inverses of B, C, and D, respectively. Then, by the triangle inequality

$$B'D' \leq B'C' + C'D'$$

with

$$B'D' = BD \cdot \frac{|\rho|}{AB \cdot AD}, \tag{3.72}$$

$$B'C' = BC \cdot \frac{|\rho|}{AB \cdot AC}, \tag{3.73}$$

$$C'D' = CD \cdot \frac{|\rho|}{AC \cdot AD}. \tag{3.74}$$

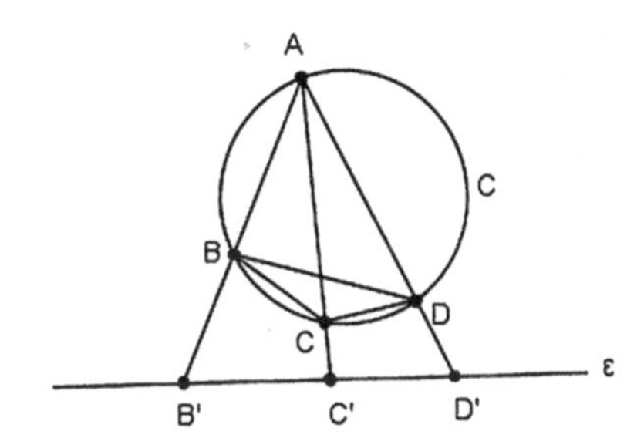

Fig. 3.38 Ptolemy's theorem (Example 3.6.2)

Hence

$$BD \cdot \frac{\rho}{AB \cdot AD} \leq (BC + CD) \cdot \frac{\rho}{AB \cdot AD},$$

which implies

$$BD \cdot AC \leq BC \cdot AD + CD \cdot AB, \tag{3.75}$$

and the assertion has been proved. □

Example 3.6.2 (Ptolemy's theorem) A quadrilateral $ABCD$ can be inscribed in a circle if and only if

$$AB \cdot CD + AD \cdot BC = AC \cdot BD, \tag{3.76}$$

i.e., (3.71) holds with equality.

Proof Let us consider the inversion of the quadrilateral $ABCD$ with pole A and power $\rho \neq 0$, with the points B', C', D' being the inverses of the points B, C, D, respectively (see Fig. 3.38). Then, the following relations hold

$$BD = B'D' \cdot \frac{|\rho|}{AB' \cdot AD'}, \tag{3.77}$$

$$AC = \frac{|\rho|}{AC'}, \tag{3.78}$$

$$BC = B'C' \cdot \frac{|\rho|}{AB' \cdot AC'}, \tag{3.79}$$

$$AD = \frac{|\rho|}{AD'}, \tag{3.80}$$

$$CD = C'D' \cdot \frac{|\rho|}{AC' \cdot AD'}, \tag{3.81}$$

$$AB = \frac{|\rho|}{AB'}. \tag{3.82}$$

Hence, by (3.71), we derive the relation

$$B'D' = B'C' + C'D', \tag{3.83}$$

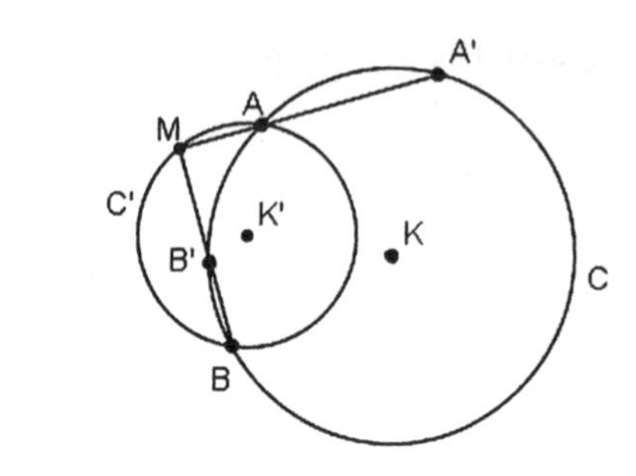

Fig. 3.39 Picture of Example 3.6.3

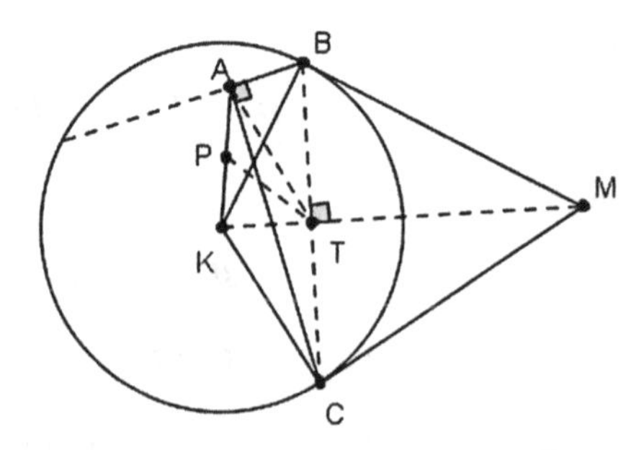

Fig. 3.40 Picture of Example 3.6.4

that is, the points B', C', D' are collinear, and the straight line that contains these points has as its inverse the circle C passing through the pole A. □

Example 3.6.3 Let C, C' be two circles with $C \cap C' = \{A, B\}$. Let $M \in C'$ and MA, MB the straight lines that intersect the other circle C at the points A', B', respectively. Prove that $A'B' \perp MK'$ with K' the center of the circle C'.

Proof Let M be the pole of an inversion with power being the inverse of the power of M with respect to the circle C (see Fig. 3.39). In this case, the inverse of the circumference C' is the straight line $A'B'$ (where A', B' are the inverse images of A, B with respect to this inversion) and thus $MK' \perp A'B'$. □

Example 3.6.4 Let a circle $C(K, r)$ and a point A in the interior of the circle be given and consider a right angle $\widehat{CAB} = 90°$, where C, B are points of the circle. If the right angle $\widehat{CAB}$ is rotated around the point A, determine the locus of the point of intersection of the tangents of $C(K, r)$ at the points C and B.

Proof Let ABC be a right triangle ($\widehat{A} = 90°$) with T the midpoint of the hypotenuse BC (see Fig. 3.40). We have

$$TA = TB = TC.$$

Hence

$$TA^2 + TK^2 = TC^2 + TK^2$$

and thus

$$TA^2 + TK^2 = KC^2 = R^2. \tag{3.84}$$

Let P be the midpoint of the straight line segment AK. By the first theorem of medians (see Chap. 4: Theorems), we get

$$TA^2 + TK^2 = 2TP^2 + \frac{AK^2}{2},$$

which yields

$$R^2 = 2TP^2 + \frac{AK^2}{2}. \tag{3.85}$$

It follows that AK is of constant length, and consequently the point T is moving on a fixed circle with center P and radius

$$r = \sqrt{\left(R^2 - \frac{AK^2}{2}\right)/2}.$$

Simultaneously, from the right triangle KBM, with $\widehat{B} = 90°$, we derive

$$KT \cdot KM = KB^2 = R^2. \tag{3.86}$$

The assertion follows. □

3.7 The Idea Behind the Construction of a Geometric Problem

To give an insight, in the present section we demonstrate the process of construction in the case of a specific problem; we consider problem *G5* from the Shortlisted Problems of the 42nd I.M.O., USA, 2001 [69].

Problem Let FBD be an acute triangle. Let EFD, ABF, and CDB be isosceles triangles exterior to FBD with

$$EF = ED, \qquad AF = AB, \quad \text{and} \quad CB = CD,$$

and such that

$$\widehat{FED} = 2\widehat{BFD},$$

$$\widehat{BAF} = 2\widehat{FBD},$$

$$\widehat{DCB} = 2\widehat{FDB}.$$

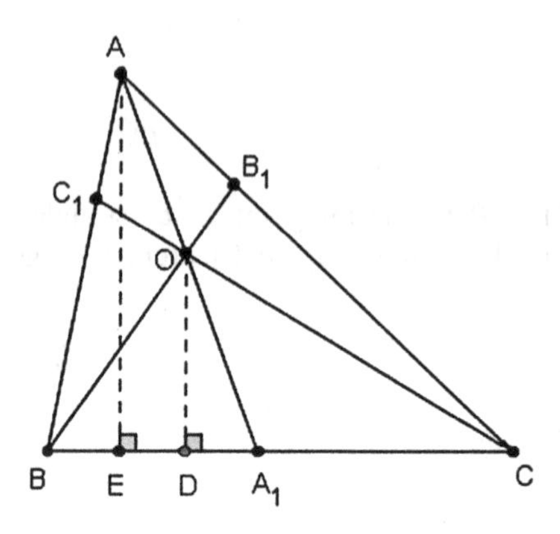

Fig. 3.41 The starting problem (Sect. 3.7)

Let

$$A_1 = AD \cap EC, \qquad C_1 = CF \cap AE, \quad \text{and} \quad E_1 = EB \cap AC.$$

Find the value of the sum

$$\frac{AD}{AA_1} + \frac{EB}{EE_1} + \frac{CF}{CC_1}.$$

It follows the construction and the solution of this problem.

1. Let ABC be a triangle, O be an interior point of the triangle, and A_1, B_1, C_1 be the points of intersection of AO, BO, and CO with the sides BC, AC, and AB, respectively. Prove that

$$\frac{OA_1}{AA_1} + \frac{OB_1}{BB_1} + \frac{OC_1}{CC_1} = 1. \tag{3.87}$$

Solution We have

$$\frac{OA_1}{AA_1} = \frac{OD}{AE} = \frac{OD \cdot BC/2}{AE \cdot BC/2} = \frac{S_{OBC}}{S_{ABC}} \tag{3.88}$$

(see Fig. 3.41). Similarly,

$$\frac{OB_1}{BB_1} = \frac{S_{OAC}}{S_{ABC}} \tag{3.89}$$

and

$$\frac{OC_1}{CC_1} = \frac{S_{OAB}}{S_{ABC}}. \tag{3.90}$$

Adding Eqs. (3.88), (3.89), and (3.90), we obtain Eq. (3.87). □

2. We consider the reflections of the point O over the sides BC, CA, and AB, respectively. We denote these points by O_1, O_2, and O_3, respectively (see Fig. 3.42). We have

$$OA_1 = A_1O_1 \quad \text{and} \quad \widehat{OA_1B} = \widehat{BA_1O_1}, \tag{3.91}$$

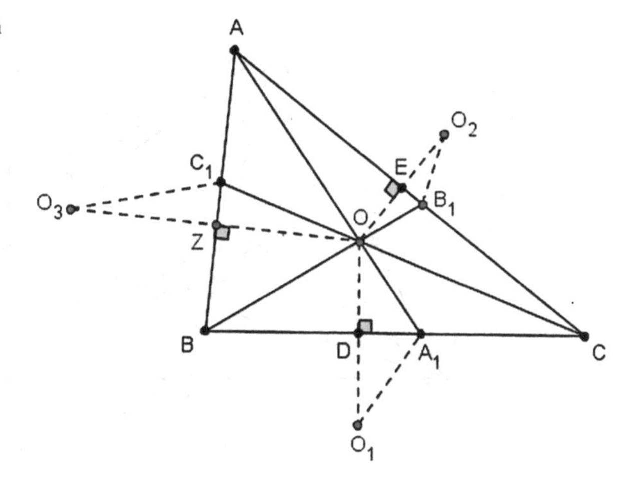

Fig. 3.42 The basic question (Sect. 3.7)

$$OB_1 = B_1O_2 \quad \text{and} \quad \widehat{OB_1A} = \widehat{AB_1O_2}, \tag{3.92}$$

and

$$OC_1 = C_1O_3 \quad \text{and} \quad \widehat{OC_1A} = \widehat{O_3C_1A}. \tag{3.93}$$

Consequently,

$$\frac{OA_1}{AA_1} = \frac{O_1A_1}{AA_1}, \tag{3.94}$$

$$\frac{OB_1}{BB_1} = \frac{O_2B_1}{BB_1}, \tag{3.95}$$

and

$$\frac{OC_1}{CC_1} = \frac{O_3C_1}{CC_1}. \tag{3.96}$$

3. We now take advantage of the equality of the angles (see Eqs. (3.91), (3.92), and (3.93)) and consider the segments AO_1, BO_2, and CO_3 which intersect the sides BC, AC, and AB at the points A_2, B_2, and C_2, respectively (see Fig. 3.43). We apply the theorem of bisectors to the triangles AO_1A_1, BO_2B_1, and CO_3C_1 and use Eq. (3.94) to obtain Eqs. (3.97), (3.98), and (3.99), which yield

$$\frac{OA_1}{AA_1} = \frac{O_1A_1}{AA_1} = \frac{O_1A_2}{A_2A}, \tag{3.97}$$

$$\frac{OB_1}{BB_1} = \frac{O_2B_1}{BB_1} = \frac{O_2B_2}{B_2B}, \tag{3.98}$$

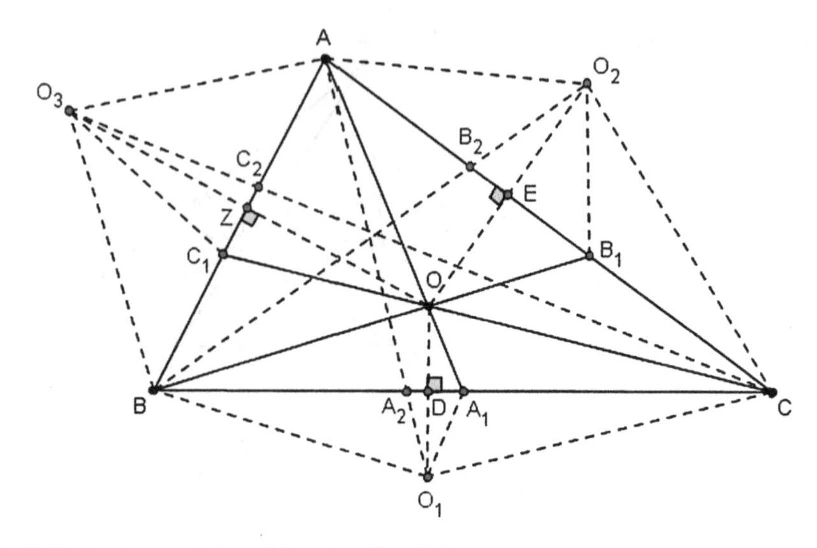

Fig. 3.43 Another translation of the ratios (Sect. 3.7)

and

$$\frac{OC_1}{CC_1} = \frac{O_3C_1}{CC_1} = \frac{O_3C_2}{C_2C}. \tag{3.99}$$

Using Eqs. (3.97), (3.98), and (3.99), we have:

$$\begin{aligned}
&\frac{AO_1}{AA_2} + \frac{BO_2}{BB_2} + \frac{CO_3}{CC_2} \\
&= \frac{AA_2 + A_2O_1}{AA_2} + \frac{BB_2 + B_2O_2}{BB_2} + \frac{CC_2 + C_2O_3}{CC_2} \\
&= \frac{AA_2}{AA_2} + \frac{BB_2}{BB_2} + \frac{CC_2}{CC_2} + \frac{O_1A_2}{AA_2} + \frac{O_2B_2}{BB_2} + \frac{O_3C_2}{CC_2} \\
&= 3 + \frac{OA_1}{AA_1} + \frac{OB_1}{BB_1} + \frac{OC_1}{CC_1} \\
&= 3 + 1 = 4.
\end{aligned}$$

Remarks

(i) Because of the fact that the point O is an interior point of the triangle ABC, the three pairs of equal angles

$$\widehat{BOC} = \widehat{CO_1B}, \tag{3.100}$$

$$\widehat{COA} = \widehat{AO_2C}, \tag{3.101}$$

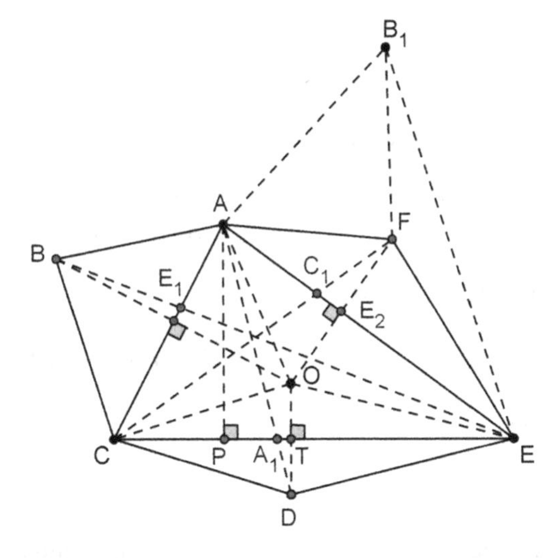

Fig. 3.44 The construction of the problem (Sect. 3.7)

$$\widehat{AOB} = \widehat{BO_3A} \tag{3.102}$$

all have measure less than π.

(ii) Evidently,

$$\begin{aligned} AO &= AO_2 = AO_3, \\ BO &= BO_1 = BO_3, \\ CO &= CO_1 = CO_2. \end{aligned} \tag{3.103}$$

(iii) The chain $AO_2CO_1BO_3A$ is a closed polygonal chain.

(iv) We have

$$\widehat{CO_1B} + \widehat{AO_2C} + \widehat{BO_3A} = 2\pi. \tag{3.104}$$

A question: Does there exist a convex hexagon $AO_2CO_1BO_3$ so that conditions (i)–(iv) hold true for the closed polygonal chain defined by it?

Answer: Yes, one can do it as long as we make sure that the reflections of O_1 over BC, of O_2 over AC, and of O_3 over AB coincide in an interior point O of ABC.

Consider the convex hexagon $ABCDEF$ that satisfies (see Fig. 3.44)

$$AB = AF,$$

$$CB = CD,$$

and

$$ED = EF,$$

and such that

$$\widehat{FED} + \widehat{BAF} + \widehat{DCB} = 2\pi. \tag{3.105}$$

Let

$$A_1 = AD \cap CE, \qquad C_1 = CF \cap AE, \quad \text{and} \quad E_1 = EB \cap AC.$$

Compute the sum

$$\frac{AD}{AA_1} + \frac{CF}{CC_1} + \frac{EB}{EE_1}. \tag{3.106}$$

Solution We have

$$\widehat{FED} + \widehat{BAF} + \widehat{DCB} = 2\pi, \tag{3.107}$$

thus

$$\widehat{CBA} + \widehat{EDC} + \widehat{AFE} = 2\pi. \tag{3.108}$$

The hexagon is convex and so all its angles are less than π. Based on the fact that

$$AF = AB, \tag{3.109}$$

we can construct outside the hexagon a triangle AFB_1 equal to the triangle ABC so that

$$FB_1 = BC \quad \text{and} \quad \widehat{AFB_1} = \widehat{CBA},$$

and thus

$$AB_1 = AC. \tag{3.110}$$

We observe that

$$\widehat{AFE} + \widehat{B_1FA} = \widehat{AFE} + \widehat{CBA}. \tag{3.111}$$

Thus

$$\begin{aligned} &\widehat{AFE} + \widehat{B_1FA} + \widehat{EDC} \\ &\quad = \widehat{AFE} + \widehat{CBA} + \widehat{EDC} = 2\pi \end{aligned} \tag{3.112}$$

with

$$\widehat{AFE} < \pi, \tag{3.113}$$

$$\widehat{CBA} < \pi, \tag{3.114}$$

and

$$\widehat{EDC} < \pi. \tag{3.115}$$

This leads to the conclusion that the point F lies in the interior of the triangle AEB_1, which is equal to the triangle ACE. Let O be the reflection of F over AE. Then O lies in the interior of the triangle ACE. Clearly,

$$OE = EF = ED \tag{3.116}$$

and

$$\widehat{FEA} = \widehat{AEO}. \tag{3.117}$$

Therefore, the point O is the reflection of D over CE. Since

$$AO = AF = AB,$$

the point O is the reflection of B over AC. We have

$$\begin{aligned} \frac{AD}{AA_1} &= 1 + \frac{A_1D}{AA_1} \\ &= 1 + \frac{DT}{AP} \\ &= 1 + \frac{OT}{AP} \\ &= 1 + \frac{S_{OCE}}{S_{ACE}}, \end{aligned} \tag{3.118}$$

where

$$AP \perp CE \quad \text{and} \quad T = OD \cap CE. \tag{3.119}$$

Similarly,

$$\frac{CF}{CC_1} = 1 + \frac{S_{OAE}}{S_{ACE}} \tag{3.120}$$

and

$$\frac{EB}{EE_1} = 1 + \frac{S_{OAC}}{S_{ACE}}. \tag{3.121}$$

Therefore,

$$\frac{AD}{AA_1} + \frac{CF}{CC_1} + \frac{EB}{EE_1} = 3 + \frac{S_{OCE} + S_{OAE} + S_{OAC}}{S_{ACE}},$$

and hence

$$\frac{AD}{AA_1} + \frac{CF}{CC_1} + \frac{EB}{EE_1} = 3 + 1 = 4. \tag{3.122}$$

□

Chapter 4
Theorems

Geometry is the most complete science.
David Hilbert (1862–1943)

In this chapter, we present some of the most essential theorems of Euclidean Geometry.

Theorem 4.1 (Thales)

- (Direct) *Let l_1, l_2 be two straight lines in the plane. Assume that l_1, l_2 intersect the four parallel, pairwise, non-coinciding, straight lines a_1, a_2, a_3, a_4 at the points A, B, C, D and A_1, B_1, C_1, D_1, respectively. Then, the equality*

$$\frac{AB}{A_1B_1} = \frac{BC}{B_1C_1} = \frac{CD}{C_1D_1} = \frac{AC}{A_1C_1} = \frac{AD}{A_1D_1}$$

  holds true.
- (Inverse) *Consider two straight lines l_1, l_2 in the plane. Let A, B, C be points on l_1 and A_1, B_1, C_1 points on l_2 such that:*

  (i) *$AA_1 \parallel CC_1$ and the points B, B_1 are in the interior of the straight line segments AC and A_1C_1, respectively, or at the exterior of the straight line segments AC and A_1C_1, respectively.*
  (ii) *The equality*

$$\frac{AB}{A_1B_1} = \frac{BC}{B_1C_1}$$

  holds true.

  Then, the parallelism relations $BB_1 \parallel AA_1$ and $BB_1 \parallel CC_1$ hold true.

Theorem 4.2 (Pythagoras) *If ABC is a right triangle with $\widehat{A} = 90°$ then*

$$BC^2 = AB^2 + AC^2,$$

or

$$a^2 = b^2 + c^2,$$

where $a = BC$, $b = AC$, and $c = AB$.

S.E. Louridas, M.Th. Rassias, *Problem-Solving and Selected Topics in Euclidean Geometry*, DOI 10.1007/978-1-4614-7273-5_4,

Theorem 4.3 (First theorem of medians) *The sum of the squares of two sides of a triangle is equal to the sum of the double of the square of the median which corresponds to the third side and the double of the square of the half of that side.*

Theorem 4.4 (Second theorem of medians) *The absolute value of the difference of the squares of two sides of a triangle is equal to the double of the product of the third side with the projection of the median (which corresponds to this side) on this side.*

Theorem 4.5 (Stewart) *Let ABC be a triangle. On the straight line BC we consider a point D. Then the relation*

$$AB^2 \cdot DC + AC^2 \cdot BD = AD^2 \cdot BC + BD \cdot DC \cdot BC,$$

holds true.

Theorem 4.6 (Angle bisectors)

- (Internal bisector) *Let AD be the internal angle bisector of the triangle ABC. Then*

$$\frac{BD}{DC} = \frac{AB}{AC}.$$

Conversely, if D is an interior point of the side BC of the triangle ABC and the relation

$$\frac{BD}{DC} = \frac{AB}{AC}$$

holds true then the straight line AD is the angle bisector of the angle $\widehat{A}$ of the triangle ABC. The equalities

$$BD = \frac{ac}{b+c} \tag{4.1}$$

and

$$DC = \frac{ab}{b+c} \tag{4.2}$$

hold true, where $a = BC$, $b = AC$, and $c = AB$.

- (External bisector) *Let AE be the external angle bisector of the triangle ABC with $AC < AB$. Then*

$$\frac{BE}{EC} = \frac{AB}{AC}.$$

Conversely, if for the external point E of the side BC the relation

$$\frac{BE}{EC} = \frac{AB}{AC}$$

holds true then the straight line AE is the angle bisector of the external angle $\pi - \widehat{A}$ *of* $\widehat{A}$ *of the triangle ABC. If* $AC < AB$ *then the relations*

$$EB = \frac{ac}{c-b}, \tag{4.3}$$

$$EC = \frac{ab}{c-b} \tag{4.4}$$

are valid, where $a = BC$, $b = AC$, *and* $c = AB$.

Note 1 The equality

$$DE = \frac{2abc}{c^2 - b^2}$$

holds.

Note 2 Using the previously mentioned relations, we also conclude that

$$\frac{BD}{DC} = \frac{BE}{EC} \neq 1 \tag{4.5}$$

holds true. In this case, we say that the points D, E are *harmonic conjugates* of the points B, C.

Theorem 4.7 (Apollonius circle) *Let the points* B, C *be given on a straight line. On this straight line we consider two points* D, E *such that*

$$\frac{BD}{DC} = \frac{BE}{EC} \neq 1.$$

The points D, E *are called* harmonic conjugates *of the points* B, C, *or alternatively, we say that the points* B, C, D, E *form a* harmonic quadruple *and we denote it by* $(B, C, D, E) = -1$.

The geometrical locus of the points M with the property:

$$\frac{MB}{MC} = \frac{BD}{DC} \neq 1$$

is the circle with diameter the straight line segment ED. This circle is called the Apollonius circle. *In the case where*

$$\frac{BD}{DC} = 1,$$

the geometrical locus of the points M with $MB = MC$ *is obviously the perpendicular bisector of BC and the harmonic conjugate of the middle point of BC is a point at infinity.*

Two basic properties of the harmonic quadruple B, C, D, E *are the following:*

- (Desargues)

$$\frac{2}{BC} = \frac{1}{BD} + \frac{1}{BE}, \quad \text{if } \frac{BD}{DC} > 1 \tag{4.6}$$

and

$$\frac{2}{BC} = \frac{1}{BD} - \frac{1}{BE}, \quad \text{if } \frac{BD}{DC} < 1. \tag{4.7}$$

- (Newton) *The relation*

$$BM^2 = MD \cdot ME$$

holds true when the point M is the midpoint of the straight line segment BC.
- *The geometrical locus of the points M of the Euclidean plane E^2 such that*

$$\frac{MB}{MC} = \frac{m}{n} \neq 1,$$

where m, n are given straight line segments, is a circle of diameter DE when the points B, C, D, E form a harmonic quadruple (*see Theorem* 4.7). *The radius R_A of this circle is given by*

$$R_A = \frac{BC \cdot MC \cdot MB}{BM^2 - MC^2}, \tag{4.8}$$

that is,

$$R_A = \frac{BC \cdot \frac{m}{n}}{(\frac{m}{n})^2 - 1}. \tag{4.9}$$

Theorem 4.8 (Vecten's point) *Let the triangle ABC be given. Consider the squares $ABDE$, $ACZH$, $BCQI$ that are externally constructed with respect to the triangle ABC. Then, the following propositions hold true* (*see Fig.* 4.1):

- $EC = HB$, $DC = AI$, $AQ = BZ$, *and* $EC \perp HB$, $AQ \perp BZ$, $DC \perp AI$.
- $EH = 2AM$ (*M is the midpoint of BC*).
- *If we consider the parallelogram $AEA'H$ then we have $\triangle EAA' = \triangle ABC$, and furthermore the median AM of the triangle ABC is an altitude of the triangle AEH and the altitude AL of the triangle ABC is the median of the triangle AEH.*
- *The straight lines BZ, CD and the altitude AL of the triangle ABC pass through the same point.*
- *The straight lines EZ, HD and the median AM of the triangle ABC pass through the same point.*
- *The circumscribed circles to the squares $ABDE$, $ACZH$ and the straight lines BH, CE, DZ, AK_1, where K_1 is the center of the square $BCQI$, pass through the same point.*

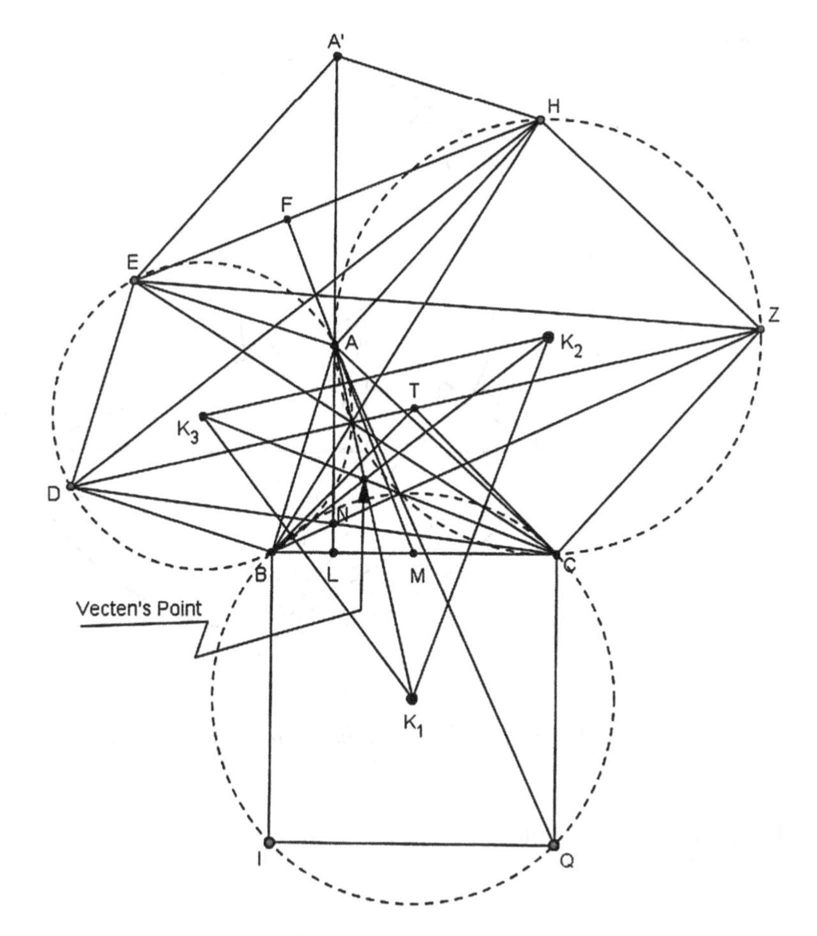

Fig. 4.1 Illustration of Theorem 4.8

- *If* K_1, K_2, K_3 *are the centers of the squares* $BCQI, ACZH, ABDE$ *then the straight lines* AK_1, BK_2, CK_3 *have a point in common, the so-called* Vecten's point. *This point is the orthocenter of the triangle* $K_1K_2K_3$.
- *If* T *is the midpoint of* DZ *then the triangle* TBC *is isosceles and orthogonal* ($\widehat{BTZ} = 90°$).

Theorem 4.9 (Euler's relation) *Let* ABC *be a triangle inscribed in a circle* (O, R). *Consider* (I, r) *to be the inscribed circle of the triangle* ABC. *Then, the relation*

$$OI^2 = R^2 - 2rR$$

holds true. The property is called Euler's relation.

We shall investigate the inverse of this proposition: Let the circles (O, R), (I, r) *be given and such that*

$$OI^2 = R^2 - 2rR.$$

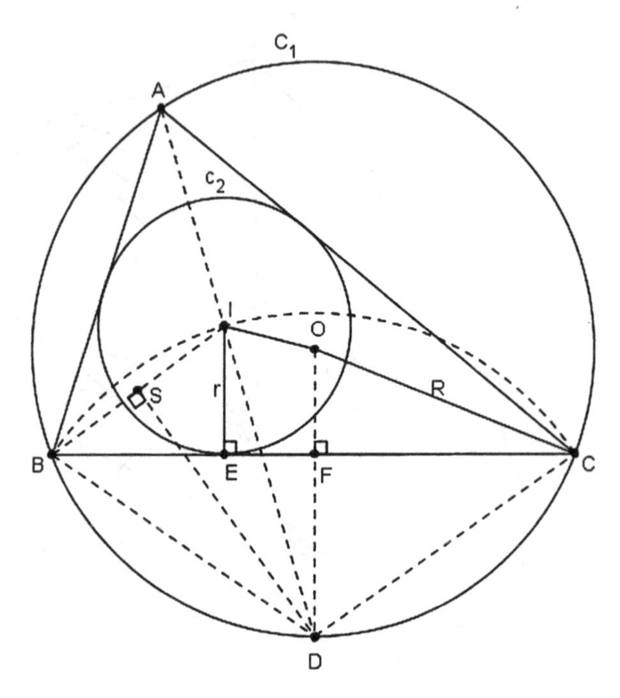

Fig. 4.2 Euler's relation (Theorem 4.9)

Then, there exists a triangle inscribed in one of the circles under consideration and circumscribed around the other one (see Fig. 4.2).

Proof Using the given condition

$$OI^2 = R^2 - 2rR,$$

we obtain

$$R \geq 2r \quad \text{and} \quad R > OI. \tag{4.10}$$

The inequalities (4.10) actually imply that the circle (I, r) is inside the circle (O, R). Let B be a point of the circle (O, R) and S be the middle point of BI. Suppose that D is the point of intersection of the perpendicular straight line to BI at S with the circle (O, R).

The perpendicular straight line to BI at the point S intersects the circle (O, R) since the point B belongs to this circle and the point I is in its interior.

The circle (D, DB) with $DB = DI$ intersects the circle (O, R) at a point C. Considering as the point A the common point of the straight line DI with the circle (O, R) it follows that I is the center of the circle inscribed in the triangle ABC.

Indeed, if r_1 is the radius of the circle inscribed in the triangle ABC, by using the well known relation of Euler and the assumption of the problem, we get

$$OI^2 = R^2 - 2Rr_1,$$

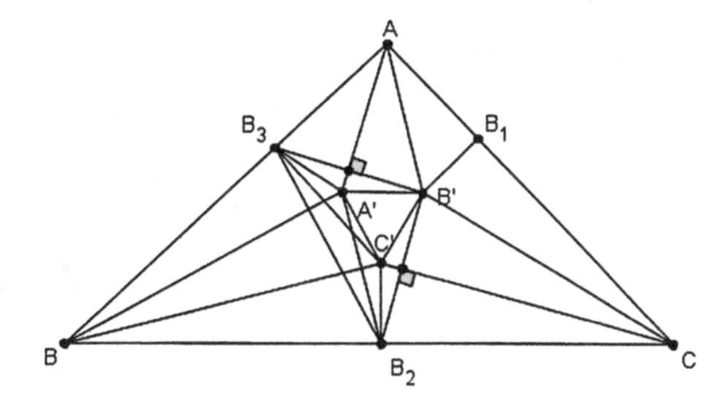

Fig. 4.3 Illustration of Morley's Theorem 4.11

and therefore,

$$R^2 - 2Rr = R^2 - 2Rr_1.$$

Hence $r = r_1$.

In conclusion, considering the point B in the circle (O, R), there is actually a triangle ABC inscribed in the circle (O, R) and such that (I, r) is the inscribed circle in the triangle ABC. Since this occurs for any choice of the point $B \in (O, R)$, we obtain an infinite family of triangles inscribed in the circle (O, R) and circumscribed around the circle (I, r). □

Theorem 4.10 *Let ABC be a triangle and P, T, R be any points on the sides BC, CA, and AB, respectively. Then the circumcircles of the triangles ART, BPR, CTP pass through a common point.*

Theorem 4.11 (Morley) *Assume ABC is a triangle. Consider the trisectors of the angles $\widehat{BAC}$ and $\widehat{CBA}$, which lie closer to the side AB of the triangle and let A' be their intersection. Similarly, let B' and C' be the corresponding intersections for the sides AC and BC, respectively. Then the triangle $A'B'C'$ is equilateral.*

Proof Let $\widehat{A} = 3a$, $\widehat{B} = 3b$, $\widehat{C} = 3c$. Let B', T, S be the points of intersection of the trisectors (see Fig. 4.3 and 4.4). If we assume that

$$2a + 2c \leq 60^\circ, \qquad 2a + 2b \leq 60^\circ,$$

and

$$2b + 2c \leq 60^\circ$$

then

$$4a + 4b + 4c \leq 180^\circ,$$

that is,

$$a + b + c \leq 45^\circ,$$

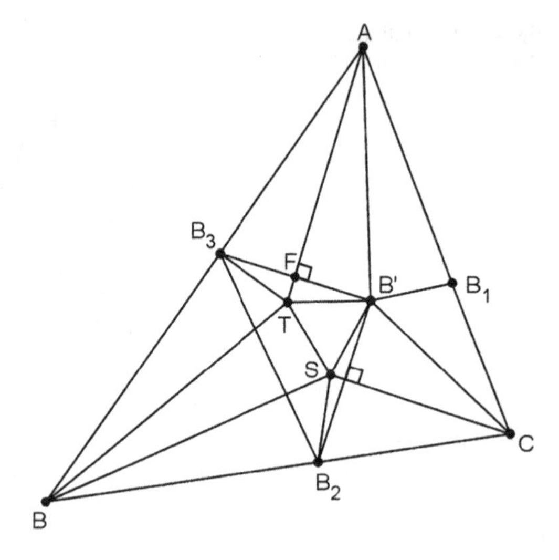

Fig. 4.4 Illustration of Morley's Theorem 4.11

which implies

$$3a+3b+3c \leq 135^\circ,$$

or equivalently,

$$180^\circ \leq 135^\circ,$$

a contradiction. We conclude that at least one of the sums $2a+2b$, $2a+2c$, $2b+2c$ should exceed 60°. Suppose that $2a+2c > 60^\circ$, then $\widehat{B} < 90^\circ$. Let

$$B'B_1 \perp AC \quad \text{and} \quad B'B_3 \perp AT.$$

Then, since every point of the angle bisector is equidistant from both sides of the angle, we get

$$B'B_3 = 2B'F = 2B'B_1.$$

Similarly, if we consider $B'B_2 \perp SC$, we obtain

$$B'B_2 = 2B'B_1,$$

and thus

$$B'B_2 = B'B_3.$$

We also observe that

$$\widehat{B_3B'B_2} = 2a+2c > 60^\circ.$$

Consider the points A', C' of the semi-straight lines AT and CS, respectively, so that the triangle $A'B'C'$ is isosceles. We obviously have

$$\widehat{A'B'B_3} = \widehat{C'B'B_2} = s,$$

therefore

$$2s = 2a + 2c - 60^\circ.$$

Observe that

$$A'B_3 = A'B' \quad \text{(due to symmetry)}$$

and

$$C'B_2 = C'B'.$$

It follows that

$$s = a + c - 30^\circ$$

and

$$2h + 2a + 2c = 180^\circ,$$

hence

$$h = \widehat{B_2B_3B'} = \widehat{B'B_2B_3}.$$

Consequently, we have

$$h = 90^\circ - a - c,$$

thus

$$\begin{aligned} h - s &= 120^\circ - 2a - 2c \\ &= 120^\circ - \frac{2}{3}(3a + 3c) \\ &= 120^\circ - \frac{2}{3}\left(180^\circ - 3b\right) \\ &= 2b, \end{aligned}$$

and thus

$$u = 2b, \tag{4.11}$$

where $u = \widehat{B_2B_3A'} = \widehat{SB_2B_3}$. At this point, we observe that

$$B_3A' = A'C' = C'B_2$$

because of the isosceles triangle $A'B'C'$, and thus

$$\begin{aligned} \widehat{B_3C'B_2} &= 180^\circ - u - \frac{u}{2} \\ &= 180^\circ - \frac{3u}{2} \\ &= 180^\circ - 3b. \end{aligned} \tag{4.12}$$

By (4.12), it follows that the quadrilateral $BB_3C'B_2$ is inscribed in a circle, and similarly, we obtain that the quadrilateral $BB_2A'B_3$ can be inscribed in a circle, as well. Consequently, the straight lines BA', BC' trisect the angle $\widehat{B}$, hence

$$T \equiv A' \quad \text{and} \quad S \equiv C'. \qquad \square$$

Theorem 4.12 (Menelaus) *Let ABC be a triangle and let D, E, F be points on the lines defined by the sides BC, CA, and AB, respectively, such that not all three of these points are interior points of the sides of the triangle. The points D, E, F are collinear if and only if the following condition holds true*

$$\frac{AF}{FB} \cdot \frac{DB}{DC} \cdot \frac{EC}{EA} = 1. \tag{4.13}$$

Theorem 4.13 (Ceva) *Let ABC be a triangle and let D, E, F be points on the sides BC, CA and AB, respectively. Then the lines AD, BE, and CF are concurrent if and only if*

$$\frac{AF}{FB} \cdot \frac{BD}{DC} \cdot \frac{CE}{EA} = 1. \tag{4.14}$$

Theorem 4.14 (Desargues) *Let two triangles ABC and DEF be given. Suppose that*

$$K = AB \cap DE, \qquad L = AC \cap DF, \quad \text{and} \quad M = BC \cap EF.$$

Then the points K, L, M are collinear if and only if the lines AD, BE, and CF are mutually parallel or concurrent.

Proof It should be enough to prove that the relation (see Fig. 4.5)

$$\frac{LD}{LF} \cdot \frac{MF}{ME} \cdot \frac{KE}{KD} = 1 \tag{4.15}$$

holds true, by applying the inverse of Menelaus' theorem to the triangle DEF. Applying Menelaus' theorem to the triangle ODF with secant the straight line LAC, we get

$$\frac{LD}{LF} \cdot \frac{CF}{CO} \cdot \frac{OA}{AD} = 1. \tag{4.16}$$

Applying Menelaus' theorem to the triangle OFE with secant the straight line MCB, we get

$$\frac{MF}{ME} \cdot \frac{BE}{BO} \cdot \frac{OC}{CF} = 1. \tag{4.17}$$

Applying again Menelaus' theorem to the triangle ODE with secant the straight line KAB, we get

$$\frac{KE}{KD} \cdot \frac{AD}{AO} \cdot \frac{BO}{BE} = 1. \tag{4.18}$$

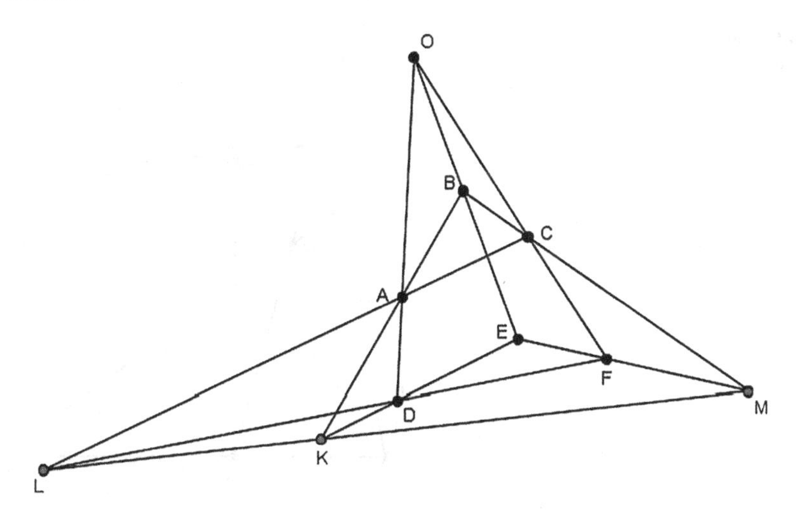

Fig. 4.5 Illustration of Desargues Theorem 4.14

Multiplying the relations (4.16)–(4.18), we finally deduce (4.15), and this completes the proof. □

Theorem 4.15 (Brahmagupta) *Consider a cyclic quadrilateral, that is, a quadrilateral whose four vertices lie on a circle, with sides of lengths a, b, c, and d. Then its area S is given by the formula*

$$S = \sqrt{(s-a)(s-b)(s-c)(s-d)},$$

where

$$s = \frac{a+b+c+d}{2}.$$

Theorem 4.16 (Simson–Wallace) *Let A, B, C be three points on a circle. Then the feet of the perpendicular lines from a point P to the lines AB, BC, CA are collinear if and only if the point P also lies on the circle* (*see Fig.* 4.6).

Theorem 4.17 (Archimedes) *Let D be the midpoint of the arc AC of a circle, B a point that lies on the arc DC, and let E be the point on AB such that DE is perpendicular to AB* (*see Fig.* 4.7). *Then*

$$AE = BE + BC.$$

Proof Let us consider the point Z on the semistraight line AB such that

$$AB < AZ \quad \text{and} \quad BZ = BC.$$

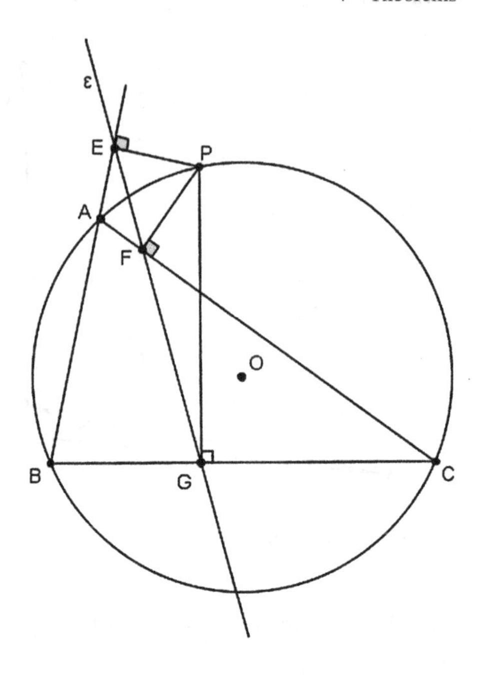

Fig. 4.6 Illustration of Simson–Wallace Theorem 4.16

Observe that

$$\begin{aligned}\widehat{DBZ} &= 180^\circ - \widehat{ABD}\\ &= 180^\circ - \widehat{ACD}\end{aligned} \tag{4.19}$$

and

$$\widehat{CBD} = 180^\circ - \widehat{DAC}. \tag{4.20}$$

The point D is the midpoint of the arc ABC, therefore

$$\widehat{ACD} = \widehat{DAC}. \tag{4.21}$$

Using the relation (4.21) and the fact that by construction $DZ = DC$, we obtain

$$\widehat{CBD} = \widehat{DBZ}. \tag{4.22}$$

By applying (4.21) and the fact that by construction $BZ = BC$, we get the equality of the triangles BCD and DZB, and thus

$$DZ = DC = DA.$$

Hence, the triangle DZA is isosceles, hence the height DE is also the median, and in conclusion,

$$AE = EZ = EB + BZ = EB + BC. \tag{4.23}$$

□

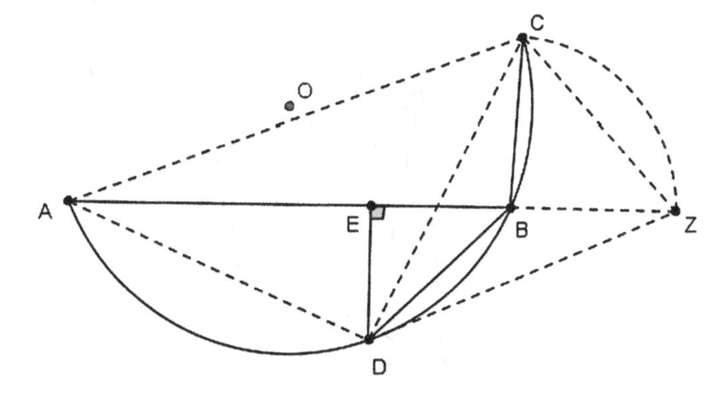

Fig. 4.7 Illustration of Archimedes Theorem 4.17

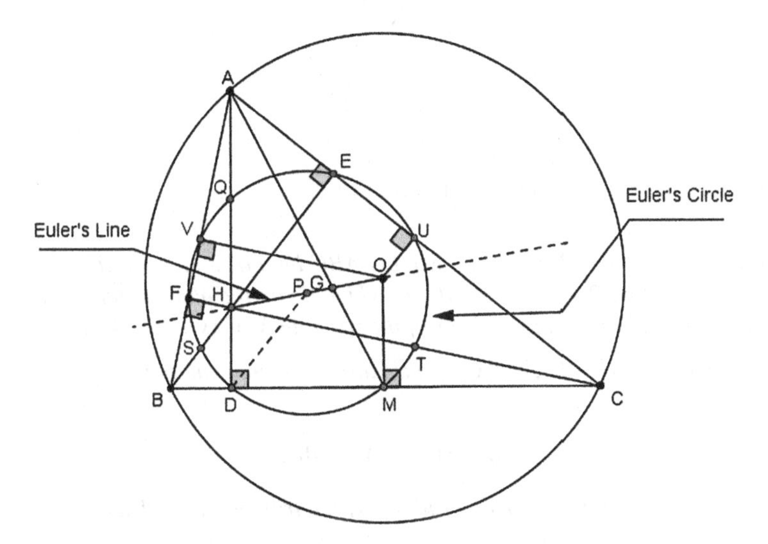

Fig. 4.8 Illustration of Euler's line and circle (Theorem 4.18)

Theorem 4.18 (Euler's line—Euler's circle)

1. *Let H be the point of intersection of the heights of the triangle ABC, O be the center of the circumscribed circle, and G be the barycenter of ABC. Then the point G lies on the segment OH and $GH = 2OG$.*

   The line that contains the points O, G, and H is called the Euler's line *of the triangle ABC.*
2. *In a triangle ABC, the midpoints of its sides, the feet of its heights, and the midpoints of the segments that connect the intersection point of the heights (orthocenter) with the vertices of ABC all lie on one circle with center of this circle being the midpoint of the straight line segment OH.*

   The above circle is called the nine-point circle *or* Euler's circle (*see Fig.* 4.8).

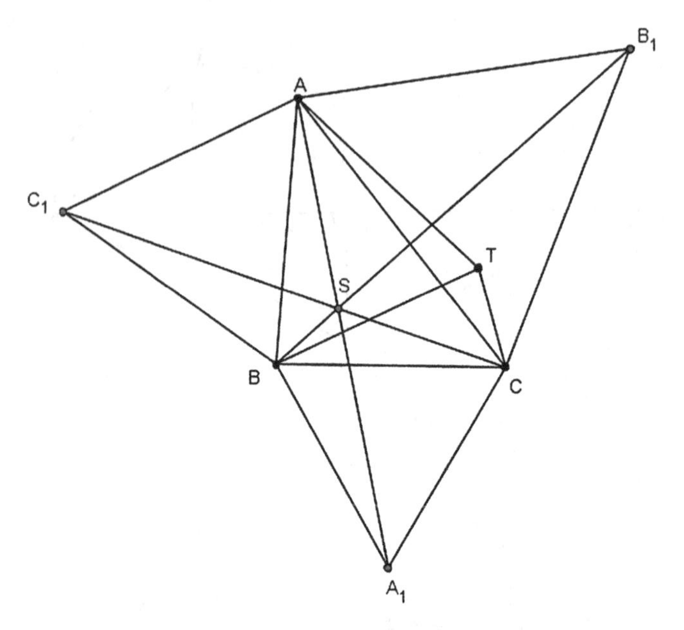

Fig. 4.9 Illustration of Fermat–Toricelli Theorem 4.19

Theorem 4.19 (Fermat–Toricelli point) *Let ABC be a triangle. Consider the equilateral triangles AC_1B, AB_1C, and BA_1C which lie on the plane determined by the points A, B, C and are in the exterior of the triangle ABC (see Fig.* 4.9).

1. *Then the lines AA_1, BB_1, and CC_1 pass through a common point S.*
2. *It holds*

$$SA + SB + SC = AA_1 = BB_1 = CC_1.$$

3. *Let T be a point in the plane determined by the triangle ABC. Then*

$$TA + TB + TC \geq SA + SB + SC,$$

when the angles of the triangle are less than 120°.

Theorem 4.20 (Miquel–Steiner) *Consider a complete quadrilateral ABCDEZ, where $E = AB \cap DC$ and $Z = AD \cap BC$. The circumscribed circles of the triangles AED, ABZ, BEC, and DCZ pass through a common point. This point is called the* Miquel's point (*see Fig.* 4.10).

Corollary 4.1 *Miquel's point belongs to the line EZ if and only if the complete quadrilateral is inscribed in a circle.*

Corollary 4.2 *In a complete quadrilateral ABCDEZ, the centers of the circumscribed circles of the triangles AED, ABZ, CBE, CZD and Miquel's point belong to the same circle.*

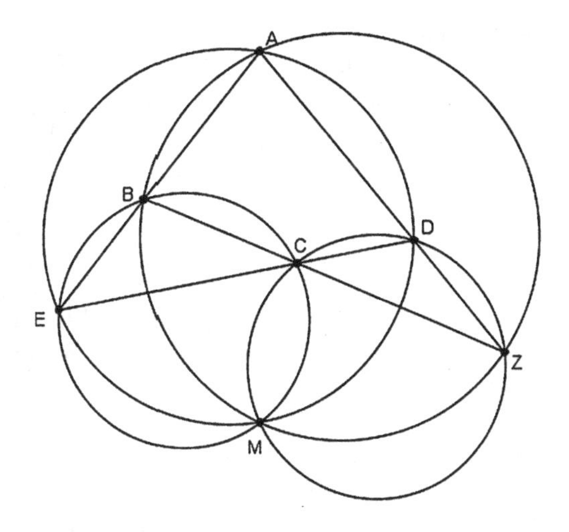

Fig. 4.10 Illustration of Miquel–Steiner Theorem 4.20

Theorem 4.21 (Feuerbach) *The Euler circle (nine-point circle) of a triangle ABC is tangent to the incircle circumference and to all the three excircles of the triangle ABC.*

Proof Let E be the point of contact of the incircle (I, r) with the side BC and Z the point of contact of the excircle (I_a, r_a) with the side BC (see Fig. 4.11). Consider the height AA', the angle bisector AD (with the points I, I_a lying on AD, as it is evident) and the midpoint M of the side BC, where BC is a common internal tangent of the circles (I, r) and (I_a, r_a).

Let LS be the other internal common tangent of the same circles. It is a fact that the point of intersection of these common tangents EZ, LS is the internal point of homothety, but also the inversion point, of the circles (I, r), (I_a, r_a), which lies on the straight line joining their centers. Consequently, it is the point D.

The straight lines BC and LS are symmetrical with respect to the straight line AD. Hence the straight line LS is anti-parallel to the straight line BC, with respect to the straight lines AB, AC.

The straight line BI is the angle bisector of the angle $\widehat{CBA}$ and BI_a is the angle bisector of its exterior angle. It follows that

$$\frac{AI}{ID} = \frac{AI_a}{I_aD}$$

and thus

$$\frac{A'E}{ED} = \frac{A'Z}{DZ}. \tag{4.24}$$

We know that

$$BE = ZC = s - b,$$

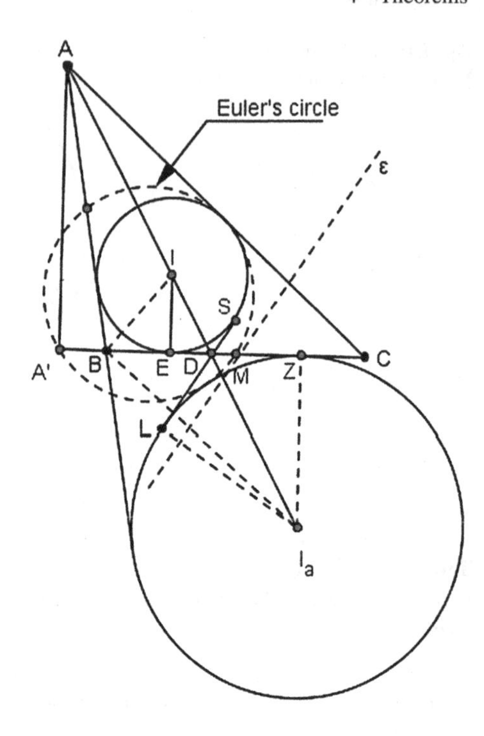

Fig. 4.11 Illustration of Feuerbach Theorem 4.21

where s is the semiperimeter of the triangle ABC, and hence

$$ME^2 = MD \cdot MA'. \tag{4.25}$$

The equality (4.25) is a necessary and sufficient condition so that

$$\frac{ED}{DZ} = \frac{A'E}{A'Z}$$

holds true, when the point M is the midpoint of the straight line segment EZ, that is, a necessary and sufficient condition so that the points A', D are harmonic conjugates of the points E, Z.

Now, if we consider the inversion with inversion pole M and power ME^2, we obtain, by means of this inversion, the transformation of the point I to the point I_a and of the point I_a to the point I. The Euler circle is transformed, through the inversion (M, ME^2), to a straight line parallel to its tangent at the point M.

We know that the Euler circle passes through the point A' and since

$$ME^2 = MD \cdot MA'$$

and the point D should belong to the inverse of the Euler circle which is a straight line parallel to the tangent of the circle at the point M, that is, the anti-parallel of the straight line BC with respect to the straight lines AB, AC. Namely, it is the straight line LS, the common internal tangent of the circles (I, r), (I_a, r_a). It follows that the straight line LS has as its inverse the Euler circle which happens to be tangent to the circles (I, r), (I_a, r_a) at the corresponding inverses of the points L, S. □

Theorem 4.22 (Brocard) *Let ABCD be a quadrilateral inscribed in a circle with center O and let*

$$E = AB \cap CD, \qquad F = AD \cap BC, \quad \text{and} \quad J = AC \cap BD.$$

Then O is the orthocenter of the triangle EFJ.

Theorem 4.23 (The butterfly theorem) *Let PQ be a chord of a circle and M be its midpoint. Let AB and CD be two other chords of the circle which pass through the point M. Let AD and BC intersect the chord PQ at the points X and Y, respectively. Then*

$$MX = MY.$$

Theorem 4.24 (Maclaurin) *Consider the angle* $\widehat{xOy}$. *Two points A and B are moving on its sides Ox and Oy, respectively, in such a way that*

$$mOA + nOB = k,$$

where m, n are given positive real numbers and k is a given straight line segment. Then the circumscribed circle of the triangle OAB passes through a fixed point.

Remark 4.1 This point lies on the straight line Oh which is the geometrical locus of the points satisfying the property that their distances from the sides of the angle $\widehat{xOy}$ are m/n.

Theorem 4.25 (Pappou–Clairaut) *Let ABC be a triangle. In the exterior of ABC, we consider the parallelograms* $ABB'A'$, $ACC'A''$, *P is the common point of the straight lines* $B'A'$, $C'A''$, *and T is the intersection point of AP, BC.*

On the extension of the straight line segment AT and in the exterior of the triangle, we consider a straight line segment TN = PA.

Let $BB''C''C$ *be the parallelogram such that the point N belongs to the side* $B''C''$. *Then the equality*

$$S_{BB''C''C} = S_{AA'B'B} + S_{ACC'A''} \tag{4.26}$$

holds true, where S denotes the enclosed area of the corresponding quadrilateral.

Observation By using the Pappou–Clairaut Theorem, one can easily prove the Pythagorean Theorem. This can be done as follows (see Fig. 4.12). We observe that the right triangles HTA and ABC are equal. Indeed, it holds

$$AC = AH \quad \text{and} \quad HT = AE = AB,$$

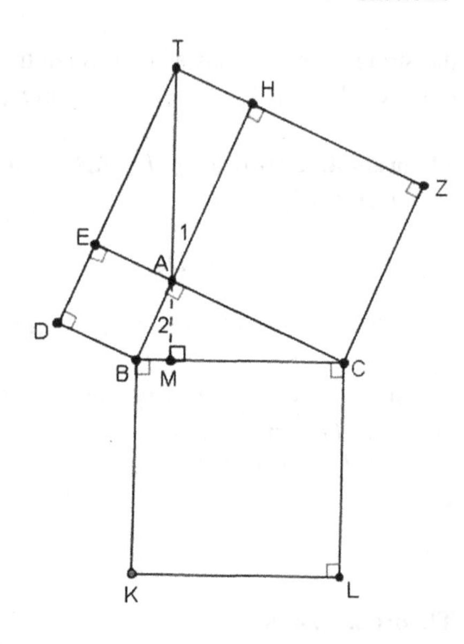

Fig. 4.12 Illustration of Pappou–Clairaut Theorem 4.25

hence

$$AT = BC = BK,$$

and therefore,

$$\widehat{BAM} = \widehat{A_1} = \widehat{C}.$$

Thus

$$\widehat{A_2} + \widehat{B} = \widehat{C} + \widehat{B} = \frac{\pi}{2},$$

where M is considered to be a point on the side BC which is the intersection point of the straight lines BC and AT. It follows that

$$AT \perp BC,$$

which implies that

$$AT \parallel BK.$$

Since the squares are also parallelograms, by a direct application of the Pappou–Clairaut Theorem, we deduce that

$$(ABDE) + (ACZH) = (BKLC),$$

that is,

$$AB^2 + AC^2 = BC^2.$$

Theorem 4.26 *Let ABC be a triangle and M be a point in its interior. If x, y, z denote the distances of the point M from the sides BC, CA, and AB, respectively, of the triangle then the product xyz attains its maximum value when the point M is identified with the barycenter of ABC.*

Theorem 4.27 (Cesáro) *Let the triangle ABC be given. Consider two triangles KLM and $K'L'M'$ circumscribed around the given triangle and similar to another given triangle with their respective homological sides perpendicular to each other, that is,*

$$KL \perp K'L', \qquad KM \perp K'M', \qquad ML \perp M'L'.$$

Then the sum of the areas of the triangles KLM and $K'L'M'$, that is,

$$S_{KLM} + S_{K'L'M'},$$

is constant, where S denotes the enclosed area of the corresponding triangle.

Theorem 4.28 *The three Apollonian circles of a given triangle form a bundle, that is, they have a chord in common.*

Theorem 4.29 *Consider a circle with center O and two points M and N in its interior such that M, N are symmetrical with respect to its center O. Let T be a point on the circle. Consider A, C, B the intersection points of the straight lines TM, TO, TN with the circle. Then, the tangent to the circumference at the point C and the straight line AB have a point in common that belongs to the straight line MN.*

Theorem 4.30 *Consider a circle and an orthogonal triangle ABC, $\widehat{A} = 90°$, inscribed in this circle. Consider the straight lines defined by the sides of the orthogonal triangle and the tangent straight lines to the circle at the points O, D, Q of the arcs BC, AB, and AC, respectively, such that*

$$OM = ON, \qquad DE = OZ, \qquad HQ = QL,$$

where

- *M, N are the intersection points of the tangent line at the point O with the straight lines AB, AC,*
- *E, Z are the intersection points of the tangent line at the point D with the straight lines AB, AC, and*
- *H, L are the intersection points of the tangent line at the point Q with the straight lines BC and AC.*

Then the triangle OQD is equilateral.

Theorem 4.31 *Let three circles be given, considered in pairs, and the six centers of similarity, three of them in the exterior and the other three in the interior. The three* *points in the exterior are collinear as well as two external points are collinear with one of the interior points.*

Theorem 4.32 *Let the triangle ABC be given. Consider the six projections of the feet of the altitudes onto the other pair of sides of the triangle. Then, the feet of these projections are homocyclic points.*

Theorem 4.33 *Let ABC be a triangle. If A_1, B_1, C_1 are the points of contact of the incircle of the triangle with the sides BC, CA, AB, respectively (or the contact points of the excircle with these sides), then the straight lines AA_1, BB_1, and CC_1 pass through the same point. This point is called the* Gergonne's point.

Theorem 4.34 (Pascal's line) *In a hexagon inscribed in a circle, the intersection points of its opposite sides are collinear.*

Theorem 4.35 (Nagel's point) *Let ABC be a triangle. If D, E, F are the contact points of the corresponding exscribed circles of the triangle with the sides bc, CA, AB then the straight lines AD, BE, and CF pass through the same point N. This point is called the* Nagel's point.

Theorem 4.36 *Let ABC be a triangle, define the* symmedian *of the triangle ABC with respect to the vertex A to be the straight semiline Ax that is the geometrical locus of the points M such that the ratio of its distances from the sides AB, AC is AB/AC. Then, the three symmedians of a triangle ABC pass through the same point which is called the* Lemoine's point.

Theorem 4.37 *Let K be the Lemoine point of the triangle ABC. Assume that the line segments KA, KB, KC are divided in analogous parts between them by using the points A_1, B_1, C_1. Then, the intersections of the straight lines B_1C_1, C_1A_1, A_1B_1 with the sides of the triangle ABC are homocyclic points (the* Tucker's circle*). The center of this circle belongs to the straight line KO, where O is the center of the circumscribed circle of the triangle ABC.*

Theorem 4.38 (Erdős–Mordell) *If from a point O situated in the interior of a given triangle ABC, we consider the perpendiculars to its sides OD, OE, OF, then*

$$OA + OB + OC \geq 2(OD + OE + OF).$$

The equality holds if and only if the triangle ABC is equilateral and O is its centroid.

Chapter 5
Problems

Problems worthy of attack prove their worth by fighting back.
Paul Erdős (1913–1996)

5.1 Geometric Problems with Basic Theory

5.1.1 Let a, b, c, d be real numbers, different from zero, such that three of them are positive and one is negative, and also

$$a+b+c+d=0.$$

Prove that there exists a triangle with sides of length

$$\sqrt{\frac{a+b}{ab}},\quad \sqrt{\frac{b+c}{bc}},\quad \sqrt{\frac{a+c}{ac}},$$

respectively.

5.1.2 Let ABC be an equilateral triangle. Find the straight line segment of minimal length such that when it moves with its endpoints sliding along the perimeter of the triangle ABC, it covers all the interior of the triangle ABC.

5.1.3 Let $PBCD$ be a rectangle inscribed in the circle (O, R). Let DP be an arc of (O, R) which does not contain the vertices of $PBCD$ and let A be a point of DP. The line parallel to DP that passes through A intersects the line BP at the point Z. Let F be the intersection point of the lines AB and DP and let Q be the intersection point of ZF and DC. Show that the straight line AQ is perpendicular to the line segment BD.

5.1.4 Let l be a straight line and H be a point not lying on l. Let S be the set of triangles that have their orthocenter at H and let ABC be one of these triangles. Let l_1, l_2, l_3 be the reflections of the line l with respect to the sides BC, CA, AB. Let $A_1 = l_2 \cap l_3$, $B_1 = l_3 \cap l_1$, and $C_1 = l_1 \cap l_2$. Show that the ratio of the perimeter of the triangle $A_1B_1C_1$ to the area of the triangle $A_1B_1C_1$ is constant.

S.E. Louridas, M.Th. Rassias, *Problem-Solving and Selected Topics in Euclidean Geometry*, DOI 10.1007/978-1-4614-7273-5_5,

5.1.5 Let ABC be a triangle. Consider two circles (K, R) and (L, r) with constant radii, which move in such a way that they remain tangent to the sides AB and AC, respectively, such that their centers belong to the interior of ABC, and finally such that the length of KL is preserved. Prove that there is a circle (M, h) (with constant radius) that moves in such a way that it remains tangent to the side BC and such that the triangle MKL has sides of constant length.

5.1.6 Let ABC and $A_1B_1C_1$ be triangles. Let AD and A_1D_1 be bisectors of the angles $\widehat{A}$ and $\widehat{A_1}$, respectively, and let CE and C_1E_1 be the distances of the vertices C, C_1 from the lines AD and A_1D_1, respectively. Suppose that

$$AD = A_1D_1,$$

$$\widehat{CBA} = \widehat{C_1B_1A_1},$$

$$CE = C_1E_1.$$

Prove that

$$ABC = A_1B_1C_1.$$

5.1.7 In the triangle ABC, let B_1, C_1 be the midpoints of the sides AC and AB, respectively, and H be the foot of the altitude passing through the vertex A.

Prove that the circumcircles of the triangles AB_1C_1, BC_1H, and B_1CH have a common point I and the line HI passes through the midpoint of the line segment B_1C_1.

(*Shortlist, 12th IMO, 1970, Budapest–Keszthely, Hungary*)

5.1.8 Let ABC be an acute triangle and AM be its median. Consider the perpendicular bisector of the side AB and let E be its common point with the median AM. Let also D be the intersection of the median AM with the perpendicular bisector of the side AC. Suppose that the point L is the intersection of the straight lines BE and CD and that L_1, L_2 are the projections of L to AC and CD, respectively. Prove that the straight line L_1L_2 is perpendicular to AM.

5.1.9 Prove that in a triangle with no angle larger than 90° the sum of the radii R, r of its circumscribed and inscribed circles, respectively, is less than the largest of its altitudes.

5.1.10 Let KLM be an equilateral triangle. Prove that there exist infinitely many equilateral triangles ABC, circumscribed to the triangle KLM such that

$$K \in AB, \quad L \in BC \quad \text{and} \quad M \in AC$$

with

$$KB = LC = MA.$$

5.1.11 Let ABC be a triangle. Consider the points

$$K \in AB, \quad L \in BC, \quad M \in AC$$

such that

$$KB = LC = MA.$$

If the triangle KLM is equilateral, prove that the same holds true for the triangle ABC.

5.1.12 Let ABC be an isosceles triangle with $\widehat{A} = 100°$. Let BL be the angle bisector of the angle $\widehat{ABC}$. Prove that

$$AL + BL = BC.$$

(Proposed by Andrei Razvan Baleanu [23], Romania)

5.1.13 Let ABC be a triangle with $\widehat{A} = 90°$ and d be a straight line passing through the incenter of the triangle and intersecting the sides AB and AC at the points P and Q, respectively. Find the minimum of the quantity $AP \cdot AQ$.

(Proposed by Dorin Andrica [17], Romania)

5.1.14 Let P be a point in the interior of a circle. Two variable perpendicular lines through P intersect the circle at the points A and B. Find the geometrical locus of the midpoint of the line segment AB.

(Proposed by Dorin Andrica [16], Romania)

5.1.15 Prove that any convex quadrilateral can be dissected into n, $n \geq 6$, cyclic quadrilaterals.

(Proposed by Dorin Andrica [19], Romania)

5.1.16 Let ABC be a triangle such that $\widehat{ABC} > \widehat{ACB}$ and let P be an exterior point in its plane such that

$$\frac{PB}{PC} = \frac{AB}{AC}.$$

Prove that

$$\widehat{ACB} + \widehat{APB} + \widehat{APC} = \widehat{ABC}.$$

(Proposed by Mircea Becheanu [25], Romania)

5.1.17 Prove that if a convex pentagon satisfies the following properties:

1. All its internal angles are equal;
2. The lengths of its sides are rational numbers,

then this is a regular pentagon.

(*18th BMO, Belgrade, Serbia*)

5.1.18 Let k points be in the interior of a square of side equal to 1. We triangulate it with vertices these k points and the square vertices. If the area of each triangle is at most $\frac{1}{12}$, prove that $k \geq 5$.

(*Proposed by George A. Tsintsifas, Greece*)

5.1.19 Let ABC be an equilateral triangle and D, E, F be points of the sides BC, CA, and AB, respectively. If the center of the inscribed circle of the triangle DEF is the center of the triangle ABC, determine what kind of triangle DEF is.

(*Proposed by George A. Tsintsifas, Greece*)

5.2 Geometric Problems with More Advanced Theory

5.2.1 Consider a circle $C(K,r)$, a point A on the circle and a point P outside the circle. A variable line l passes through the point P and intersects the circle at the points B and C. Let H be the orthocenter of the triangle ABC. Prove that there exists a unique point T in the plane of the circle $C(K,r)$ such that the sum

$$HA^2 + HT^2$$

remains constant (independent of the position of the line l).

5.2.2 Consider two triangles ABC and $A_1B_1C_1$ such that

1. The lengths of the sides of the triangle ABC are positive consecutive integers and the same property holds for the sides of the triangle $A_1B_1C_1$.
2. The triangle ABC has an angle that is twice the measure of one of its other angles and the same property holds for the triangle $A_1B_1C_1$.

Compare the areas of the triangles ABC and $A_1B_1C_1$.

5.2.3 Let a triangle ABC be given. Investigate the possibility of determining a point M in the interior of ABC such that if D, E, Z are the projections of M to the sides AB, BC, CA, respectively, then the relations

$$\frac{AD}{m} = \frac{BE}{n} = \frac{CZ}{l}$$

should hold if m, n, and l are the lengths of given line segments.

5.2.4 Let $\widehat{xOy}$ be a right angle and on the side Ox fix two points A, B with $OA < OB$. On the side Oy, we consider two moving points C, D such that $OD < OC$ with $CD/DO = m/n$, where m, n are given positive integers. If M is the point of intersection of AC and BD, determine the position of M under the assumption that the angle $\widehat{DMA}$ attains its minimum.

5.2.5 Given $\widehat{xOy} = 60^\circ$, we consider the points A, B moving on the sides Ox and Oy, respectively, so that the length of the line segment AB is preserved subject to the assumption that the triangle OAB is not an obtuse triangle. Let D, E, Z be the feet of the heights OD, AE, and BZ of the triangle OAB to AB, BO, and OA, respectively. Compute the maximal value of the sum

$$\sqrt{DE} + \sqrt{EZ} + \sqrt{ZD}.$$

5.2.6 Let O be a given point outside a given circle of center C. Let OPQ be any secant of the circle passing through O and R be a point on PQ such that

$$\frac{OP}{QO} = \frac{PR}{RQ}.$$

Find the geometrical locus of the point R.

5.2.7 Prove that in each triangle the following equality holds:

$$\frac{1}{r}\left(\frac{b^2}{r_b} + \frac{c^2}{r_c}\right) - \frac{a^2}{r_b r_c} = 4\left(\frac{R}{r_a} + 1\right),$$

where s is the semiperimeter of the triangle, S is the area enclosed by the triangle, a, b, c are the sides of the triangle, R is the radius of the circumscribed circle, r is the corresponding radius of the inscribed circle, and r_a, r_b, r_c are the radii of the corresponding exscribed circles of the triangle.

(*Proposed by Dorin Andrica, Romania, and Khoa Lu Nguyen [14], USA*)

5.2.8 Let $A_1A_2A_3A_4A_5$ be a convex planar pentagon and let $X \in A_1A_2$, $Y \in A_2A_3$, $Z \in A_3A_4$, $U \in A_4A_5$, and $V \in A_5A_1$ be points such that A_1Z, A_2U, A_3V, A_4X, A_5Y intersect at the point P. Prove that

$$\frac{A_1X}{A_2X} \cdot \frac{A_2Y}{A_3Y} \cdot \frac{A_3Z}{A_4Z} \cdot \frac{A_4U}{A_5U} \cdot \frac{A_5V}{A_1V} = 1.$$

(*Proposed by Ivan Borsenko [26], USA*)

5.2.9 Given an angle $\widehat{xOy}$ and a point S in its interior, consider a straight line passing through S and intersecting the sides Ox, Oy at the points A and B, respectively. Determine the position of AB so that the product $OA \cdot OB$ attains its minimum.

5.2.10 Let the incircle of a triangle ABC touch the sides BC, CA, AB at the points D, E, F, respectively. Let K be a point on the side BC and M be the point on the line segment AK such that $AM = AE = AF$. Denote by L and N the incenters of the triangles ABK and ACK, respectively.

Prove that K is the foot of the altitude from A if and only if $DLMN$ is a square.

(*Proposed by Bogdan Enescu [41], Romania*)

5.2.11 Let $ABCD$ be a square of center O. The parallel through O to AD intersects AB and CD at the points M and N, respectively, and a parallel to AB intersects the diagonal AC at the point P. Prove that

$$OP^4 + \left(\frac{MN}{2}\right)^4 = MP^2 \cdot NP^2.$$

(Proposed by Titu Andreescu [7], USA)

5.2.12 Let O, I, H be the circumcenter, the incenter, and the orthocenter of the triangle ABC, respectively, and let D be a point in the interior of ABC such that

$$BC \cdot DA = CA \cdot DB = AB \cdot DC.$$

Prove that the points A, B, D, O, I, H are cocyclic if and only if $\widehat{C} = 60°$.

(Proposed by T. Andreescu, USA, and D. Andrica and C. Barbu [8], Romania)

5.2.13 Let H be the orthocenter of an acute triangle ABC and let A', B', C' be the midpoints of the sides BC, CA, AB, respectively. Denote by A_1 and A_2 the intersections of the circle $(A', A'H)$ with the side BC. In the same way, we define the points B_1, B_2 and C_1, C_2, respectively. Prove that the points A_1, A_2, B_1, B_2, C_1, C_2 are cocyclic.

(Proposed by Catalin Barbu [24], Romania)

5.2.14 Let ABC be a triangle with midpoints M_a, M_b, M_c and let X, Y, Z be the points of tangency of the incircle of the triangle $M_aM_bM_c$ with M_bM_c, M_cM_a, and M_aM_b, respectively.

(a) Prove that the lines AX, BY, CZ are concurrent at some point P.
(b) If A_1, B_1, C_1 are points of the sides BC, AC, AB, respectively, such that the straight lines AA_1, BB_1, CC_1 are concurrent at the point P, then the perimeter of the triangle $A_1B_1C_1$ is greater than or equal to the semiperimeter of the triangle ABC.

(Proposed by Roberto Bosch Cabrera [34], Cuba)

5.2.15 Let I_a be the excenter corresponding to the side BC of a triangle ABC. Let A', B', C' be the tangency points of the excircle of center I_a with the sides BC, CA, and AB, respectively. Prove that the circumcircles of the triangles AI_aA', BI_bB', CI_aC' have a common point, different from I_a, situated on the line G_aI_a, where G_a is the centroid of the triangle $A'B'C'$.

(Proposed by Dorin Andrica [20], Romania)

5.2.16 Let C_1, C_2, C_3 be concentric circles with radii $R_1 = 1$, $R_2 = 2$, and $R_3 = 3$, respectively. Consider a triangle ABC with $A \in C_1$, $B \in C_2$, and $C \in C_3$. Prove that

$$\max S_{ABC} < 5,$$

where max S_{ABC} denotes the greatest possible area attained by the triangle ABC.
(Proposed by Roberto Bosch Cabrera [35], Cuba)

5.2.17 Consider an angle $\widehat{xOy} = 60^\circ$ and two points A, B moving on the sides Ox, Oy, respectively, so that $AB = a$, where a is a given straight line segment. Let AD, BE be the angle bisectors of $\widehat{A}$, $\widehat{B}$ in the triangle OAB. Determine the position for which the product

$$AE^m \cdot BD^n$$

attains its maximum value, when m, n are positive rational numbers expressing the lengths of two straight line segments.

5.2.18 Let $\widehat{xOy} = 90^\circ$ and points $A \in Ox$, $B \in Oy$ (with $A \neq O$, $B \neq O$), so that the condition

$$OA + OB = 2\lambda$$

holds, where $\lambda > 0$ is a given positive number. Prove that there exists a unique point $T \neq O$ such that

$$S_{OATB} = \lambda^2,$$

independently of the position of the straight line segment AB.

5.2.19 Let a given quadrilateral $A'B'C'D'$ be inscribed in a circle (O, R). Consider a straight line y intersecting the straight lines $A'D'$, $B'C'$, $B'A'$, and $D'C'$, at the points A, A_1, B, B_1, respectively, and also the circle (O, R) at the points M, M_1. Prove that

$$\sqrt{MA \cdot MA_1 \cdot MB \cdot MB_1} + \sqrt{M_1A \cdot M_1A_1 \cdot M_1B \cdot M_1B_1}$$
$$= \sqrt{(MA \cdot MA_1 + M_1A \cdot M_1A_1) \cdot (MB \cdot MB_1 + M_1B \cdot M_1B_1)}.$$

5.2.20 Let ABC be a triangle with $\widehat{BCA} = 90^\circ$ and let D be the foot of the altitude from the vertex C. Let X be a point in the interior of the segment CD. Let K be the point on the segment AX, such that $BK = BC$. Similarly, let L be the point on the segment BX such that $AL = AC$. Let M be the point of intersection of AL and BK. Show that $MK = ML$.
(53rd IMO, 2012, Mar del Plata, Argentina)

5.2.21 Let AB be a straight line segment and C be a point in its interior. Let $C_1(O, r)$, $C_2(K, R)$ be two circles passing through A, B and intersecting each other orthogonally. If the straight line DC intersects the circle C_2 at the point M, compute the supremum of $x \in \mathbb{R}$, where

$$x = S_{MAC}$$

denotes the area of the triangle MAC.

5.2.22 Let $ABCWD$ be a pentagon inscribed in a circle of center O. Suppose that the center O is located in the common part of the triangles ACD and BCW, where the point W is the intersection of the height of the triangle ACD, passing through the vertex A, with the circle. Let E be the intersection point of the straight line OK with the straight line AW, where K is the midpoint of the side AD. Suppose that the diagonal BW passes through the point E. Let Q be the common point of the diagonal BW with the straight line OK such that $ZQ \parallel AW$ and let Z be the point of intersection of the diagonals AC and BW. Compute the sum

$$\widehat{CDB} + \widehat{CBA}.$$

5.2.23 On the straight line ϵ consider the collinear points A, B, C and let $AB > BC$. Construct the semicircumferences (O_1), (O_2) with diameters AB, BC, respectively, and let D, E be their intersection points with the semicircle (O) having as diameter the line segment O_1O_2. Define the points

$$D' \equiv (O_1) \cap DE, \qquad E' \equiv (O_2) \cap DE.$$

Prove that the points

$$P \equiv AD' \cap CE', \qquad Q \equiv AD \cap CE$$

and the midpoint M of the straight line segment AC are collinear.
(*Proposed by Kostas Vittas, Greece*)

5.2.24 Let $\widehat{xOy}$ be an angle and A, B points in the interior of $\widehat{xOy}$. Investigate the problem of the constructibility of a point $C \in Ox$ in such a manner that

$$OD \cdot OE = OC^2 - CD^2, \tag{5.1}$$

where

$$D \equiv CA \cap Oy \quad \text{and} \quad E \equiv CB \cap Oy.$$

5.2.25 Let ABC be a triangle satisfying the following property: there exists an interior point L such that

$$\widehat{LBA} = \widehat{LCA} = 2\widehat{B} + 2\widehat{C} - 270°.$$

Let B', C' be the symmetric points of the points B and C with respect to the straight lines AC and AB, respectively. Prove that

$$AL \perp C'B'.$$

5.2.26 Let $AB = a$ be a straight line segment. On its extension towards the point B, consider a point C such that $BC = b$. With diameter the straight line segments AB and AC, we construct two semicircumferences on the same side of the straight line AC. The perpendicular bisector to the straight line segment BC intersects the exterior semicircumference at a point E. Prove or disprove the following assertion 1 and solve problem 2:

1. There exists a circle inscribed in the curved triangle $ABEA$.
2. If K is the center of the previously inscribed circle and M is the point of intersection of the straight line BK with the semicircumference of diameter AC, compute the area of the domain that is bounded from the semicircumference of diameter AC and the perimeter of the triangle MAC.

5.2.27 Let ABC be a triangle with $AB \geq BC$. Consider the point M on the side BC and the isosceles triangle KAM with $KA = KM$. Let the angle $\widehat{AKM}$ be given such that the points K, B are in different sides of the straight line AM satisfying the condition

$$360^\circ - 2\widehat{B} > \widehat{AKM} > 2\widehat{C}.$$

The circle (K, KA) intersects the sides AB, AC at the points D and E, respectively. Find the position of the point $M \in BC$ so that the area of the quadrilateral $ADME$ attains its maximum value.

5.2.28 Let $ABCD$ be a cyclic quadrilateral, $AC = e$ and $BD = f$. Let us denote by r_a, r_b, r_c, r_d the radii of the incircles of the triangles BCD, CDA, DAB, and ABC, respectively. Prove the following equality

$$e \cdot r_a \cdot r_c = f \cdot r_b \cdot r_d. \tag{5.2}$$

(*Proposed by Nicuşor Minculete and Cătălin Barbu, Romania*)

5.2.29 Prove that for any triangle the following equality holds

$$-\frac{a^2}{r} + \frac{b^2}{r_c} + \frac{c^2}{r_b} = 4R - 4r_a, \tag{5.3}$$

where a, b, c are the sides of the triangle, R is the radius of the circumscribed circle, r is the corresponding radius of the inscribed circle, and r_a, r_b, r_c are the radii of the corresponding exscribed circles of the triangle.

(*Proposed by Nicuşor Minculete and Cătălin Barbu, Romania*)

5.2.30 For the triangle ABC let $(x, y)_{ABC}$ denote the straight line intersecting the union of the straight line segments AB and BC at the point X and the straight line segment AC at the point Y in such a way that the following relation holds

$$\frac{\widetilde{AX}}{AB + BC} = \frac{AY}{AC} = \frac{xAB + yBC}{(x + y)(AB + BC)},$$

where $\widetilde{AX}$ is either the length of the line segment AX in case X lies between the points A, B, or the sum of the lengths of the straight line segments AB and BX

if the point X lies between B and C. Prove that the three straight lines $(x, y)_{ABC}$, $(x, y)_{BCA}$, and $(x, y)_{CBA}$ concur at a point which divides the straight line segment NI in a ratio $x : y$, where N is Nagel's point and I the incenter of the triangle ABC.

(Proposed by Todor Yalamov, Sofia University, Bulgaria)

5.2.31 Let T be the Torricelli's point of the convex polygon $A_1A_2 \ldots A_n$ and (d) a straight line such that $T \in (d)$ and $A_k \notin (d)$, where $k = 1, 2, \ldots, n$. If we denote by $B_1, B_2, \ldots, B_n$ the projections of the vertices $A_1, A_2, \ldots, A_n$ on the line (d), respectively, prove that

$$\sum_{k=1}^{n} \frac{\overrightarrow{TB_k}}{TA_k} = \vec{0}.$$

(Proposed by Mihály Bencze, Braşov, Romania)

5.2.32 Let $ABCD$ be a quadrilateral. We denote by E the midpoint of the side AB, F the centroid of the triangle ABC, K the centroid of the triangle BCD, and G the centroid of the given quadrilateral. For all points M of the plane of the quadrilateral, different from A, E, F, G, prove the following inequality:

$$\frac{6MB}{MA \cdot ME} + \frac{2MC}{ME \cdot MF} + \frac{MD}{MF \cdot MG} \geq \frac{5MK}{MA \cdot MG}.$$

(Proposed by Mihály Bencze, Braşov, Romania)

5.2.33 Let the angle $\widehat{xOy}$ be given and let A be a point in its interior. Construct a triangle ABC with $B \in Ox$, $C \in Oy$, $\widehat{BAC} = \widehat{\omega}$ such that $AB \cdot AC = k^2$, where k is the length of a given straight line segment and $\widehat{\omega}$ is a given angle.

5.2.34 Let a triangle ABC with $BC = a$, $AC = b$, $AB = c$ and a point D in the interior of the side BC be given. Let E be the harmonic conjugate of D with respect to the points B and C. Determine the geometrical locus of the center of the circumferences DEA when D is moving along the side BC.

5.3 Geometric Inequalities

5.3.1 Consider the triangle ABC and let H_1, H_2, H_3 be the intersection points of the altitudes AA_1, BB_1, CC_1, with the circumscribed circle of the triangle ABC, respectively. Show that

$$\frac{H_2H_3^2}{BC^2} + \frac{H_3H_1^2}{CA^2} + \frac{H_1H_2^2}{AB^2} \geq 3.$$

5.3.2 Let ABC be a triangle with $AB = c$, $BC = a$ and $CA = b$ and let d_a, d_b, d_c be its internal angle bisectors. Show that

$$\frac{1}{d_a} + \frac{1}{d_b} + \frac{1}{d_c} > \frac{1}{a} + \frac{1}{b} + \frac{1}{c}.$$

5.3.3 Let ABC be a triangle with $\widehat{C} > 10°$ and $\widehat{B} = \widehat{C} + 10°$. Consider a point E on AB such that $\widehat{ACE} = 10°$ and let D be a point on AC such that $\widehat{DBA} = 10°$. Let $Z \neq A$ be a point of intersection of the circumscribed circles of the triangles ABD and AEC. Show that $\widehat{ZBA} > \widehat{ZCA}$.

5.3.4 Let ABC be a triangle of area S and D, E, F be points on the lines BC, CA, and AB, respectively. Suppose that the perpendicular lines at the points D, E, F to the lines BC, CA, and AB, respectively, intersect the circumcircle of ABC at the pairs of points (D_1, D_2), (E_1, E_2), and (F_1, F_2), respectively. Prove that

$$\begin{aligned}&|D_1B \cdot D_1C - D_2B \cdot D_2C| \\ &\quad + |E_1C \cdot E_1A - E_2C \cdot E_2A| + |F_1A \cdot F_1B - F_2A \cdot F_2B| > 4S.\end{aligned}$$

5.3.5 Let ABC be an equilateral triangle and let D, E be points on its sides AB and AC, respectively. Let F, G be points on the segments AE and AD, respectively, such that the lines DF and EG bisect the angles $\widehat{EDA}$ and $\widehat{AED}$, respectively. Prove that

$$S_{DEF} + S_{DEG} \leq S_{ABC}.$$

When does the equality hold?

5.3.6 Let PQR be a triangle. Prove that

$$\frac{1}{y+z-x} + \frac{1}{z+x-y} + \frac{1}{x+y-z} \geq \frac{1}{x} + \frac{1}{y} + \frac{1}{z},$$

where

$$x = \sqrt{\sqrt[3]{QR^2} + \sqrt[5]{QR^2}}, \qquad y = \sqrt{\sqrt[3]{PR^2} + \sqrt[5]{PR^2}}, \quad \text{and} \quad z = \sqrt{\sqrt[3]{PQ^2} + \sqrt[5]{PQ^2}}.$$

5.3.7 The point O is considered inside the convex quadrilateral $ABCD$ of area S. Suppose that K, L, M, N are interior points of the sides AB, BC, CD, and DA, respectively. If $OKBL$ and $OMDN$ are parallelograms of areas S_1 and S_2, respectively, prove that

$$\sqrt{S_1} + \sqrt{S_2} < 1.25\sqrt{S},$$
$$\sqrt{S_1} + \sqrt{S_2} < C_0\sqrt{S},$$

where

$$C_0 = \max_{0<\alpha<\frac{\pi}{4}} \frac{\sin(2\alpha + \frac{\pi}{4})}{\cos\alpha}.$$

(*Proposed by Nairi Sedrakyan [88], Armenia*)

5.3.8 Let $ABCD$ be a quadrilateral with $\widehat{A} \geq 60°$. Prove that

$$AC^2 \leq 2(BC^2 + CD^2),$$

with equality, when $AB = AC$, $BC = CD$ and $\widehat{A} = 60°$.
(*Proposed by Titu Andreescu [6], USA*)

5.3.9 Let R and r be the circumradius and the inradius of the triangle ABC with sides of lengths a, b, c. Prove that

$$2 - 2\sum_{cycl}\left(\frac{a}{b+c}\right)^2 \leq \frac{r}{R}.$$

(*Proposed by Dorin Andrica [18], Romania*)

5.3.10 Let $A_1A_2\ldots A_n$ be a regular n-gon inscribed in a circle of center O and radius R. Prove that for each point M in the plane of the n-gon the following inequality holds

$$\prod_{k=1}^{n} MA_k \leq \left(OM^2 + R^2\right)^{n/2}.$$

(*Proposed by Dorin Andrica [15], Romania*)

5.3.11 Let (K_1, a), (K_2, b), (K_3, c), (K_4, d) be four cyclic disks of a plane Π, having at least one common point. Let I be a point of their intersection. Let also O be a point in the plane Π such that

$$\min\{(OA), (OA'), (OB), (OB'), (OC), (OC'), (OD), (OD')\} \geq (OI) + 2\sqrt{2},$$

where AA', BB', CC', DD' are the diameters of (K_1, a), (K_2, b), (K_3, c), and (K_4, d), respectively. Prove that

$$\begin{aligned}
&144 \cdot \left(a^4 + b^4 + c^4 + d^4\right) \cdot \left(a^8 + b^8 + c^8 + d^8\right) \\
&\geq \left[\left(\frac{ab+cd}{2}\right)^2 + \left(\frac{ad+bc}{2}\right)^2 + \left(\frac{ac+bd}{2}\right)^2\right] \\
&\quad \cdot \left[(a+b)\cdot(c+d) + (a+d)\cdot(b+c) + (a+c)\cdot(b+d)\right].
\end{aligned}$$

Under what conditions does the equality hold?

5.3.12 Let the circle (O, R) be given and a point A on this circle. Consider successively the arcs AB, BD, DC such that

$$\operatorname{arc} AB < \operatorname{arc} AD < \operatorname{arc} AC < 2\pi.$$

Using the center K of the arc BD, the center L of BD, and the corresponding radii, we draw circles that intersect the straight semilines AB, AC at the points Z and E, respectively. If

$$A' \equiv AL \cap DC, \qquad K' \equiv AK \cap BD,$$

prove that

$$\frac{3}{4}(AB \cdot AZ + AC \cdot AE) < 2R^2 + \frac{R(AK' + AL')}{2} + \frac{AB^2 + AC^2}{4}.$$

Is this inequality the best possible?

Chapter 6
Solutions

You are never sure whether or not a problem is good unless you actually solve it.
Mikhail Gromov (Abel Prize, 2009)

6.1 Geometric Problems with Basic Theory

6.1.1 Let a, b, c, d be real numbers, different from zero, such that three of them are positive and one is negative, and furthermore

$$a + b + c + d = 0. \tag{6.1}$$

Prove that there exists a triangle with sides of length

$$\sqrt{\frac{a+b}{ab}}, \quad \sqrt{\frac{b+c}{bc}}, \quad \sqrt{\frac{a+c}{ac}}, \tag{6.2}$$

respectively.

Solution If we consider

$$\frac{1}{a} = x, \qquad \frac{1}{b} = y, \qquad \frac{1}{c} = z, \quad \text{and} \quad \frac{1}{d} = u,$$

the problem assumes the following form: *Let x, y, z, u be real numbers such that three of them are positive and the condition*

$$\frac{1}{x} + \frac{1}{y} + \frac{1}{z} + \frac{1}{u} = 0$$

holds true. Prove that there exists a triangle with sides of lengths

$$\sqrt{x+y}, \quad \sqrt{y+z}, \quad \sqrt{z+x}.$$

Assume that

$$x < 0, \quad y > 0, \quad z > 0, \quad \text{and} \quad u > 0.$$

S.E. Louridas, M.Th. Rassias, *Problem-Solving and Selected Topics in Euclidean Geometry*, DOI 10.1007/978-1-4614-7273-5_6,

It is then true that

$$-\frac{1}{x}=\frac{1}{y}+\frac{1}{z}+\frac{1}{u}>\frac{1}{y}. \tag{6.3}$$

Therefore,

$$|x|<y, \tag{6.4}$$

and thus

$$x+y>0. \tag{6.5}$$

Similarly,

$$z+x>0. \tag{6.6}$$

It follows that

$$\sqrt{y+z}>\sqrt{x+y} \quad \text{and} \quad \sqrt{y+z}>\sqrt{z+x}.$$

We have

$$\begin{aligned} &\sqrt{x+y}+\sqrt{z+x}>\sqrt{y+z} \\ \Leftrightarrow\quad &\sqrt{x+y}\cdot\sqrt{z+x}>-x \\ \Leftrightarrow\quad &(z+x)\cdot(y+x)>x^2 \\ \Leftrightarrow\quad &xy+yz+zx>0. \end{aligned}$$

Since $xyz<0$ with

$$-\frac{1}{u}=\frac{1}{x}+\frac{1}{y}+\frac{1}{z}=\frac{xy+yz+zx}{xyz}<0,$$

we get that

$$xy+yz+zx>0.$$

Thus

$$\sqrt{x+y}+\sqrt{z+x}>\sqrt{y+z}. \tag{6.7}$$

Furthermore,

$$|\sqrt{x+y}-\sqrt{z+x}|<\sqrt{y+z}. \tag{6.8}$$

To verify Eq. (6.8), we write

$$|\sqrt{x+y}-\sqrt{z+x}|<\sqrt{y+z},$$

that is,

$$x<\sqrt{(x+y)(z+x)},$$

where the last inequality holds true. From Eqs. (6.7) and (6.8), the result follows. □

Fig. 6.1 Illustration of Problem 6.1.2

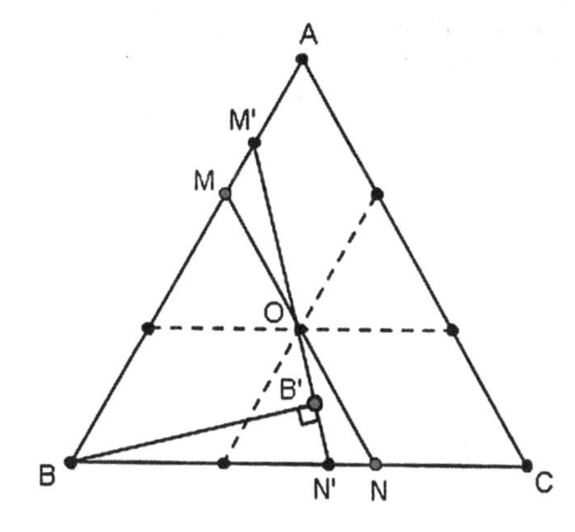

6.1.2 Let ABC be an equilateral triangle. Find the straight line segment of minimal length such that when it moves with its endpoints sliding along the perimeter of the triangle ABC, it covers all the interior of the triangle ABC.

Solution Since the straight line segment will cover all the interior of the triangle ABC, it will pass through its barycenter O as well. We shall prove that among all segments that pass through the barycenter of the triangle ABC and have their end-points on the sides of the triangle, the straight line segment with the smallest length is the segment MN which is parallel to AC. Let $M'N'$ be a straight line segment with endpoints on the sides of the triangle ABC and which passes through O (see Fig. 6.1). We have

$$OM' > OM = ON > ON' \tag{6.9}$$

and

$$\widehat{OMM'} = 120^\circ \quad \text{and} \quad \widehat{N'NO} = 60^\circ. \tag{6.10}$$

At the same time

$$\widehat{MOM'} = \widehat{NON'}, \tag{6.11}$$

therefore, there will exist an interior point T such that the triangle OMT is equal to the triangle $ON'N$. Hence

$$S_{OMM'} > S_{ONN'}. \tag{6.12}$$

Hence

$$S_{BM'N'} > S_{BMN}, \tag{6.13}$$

and thus

$$M'N' \cdot BB' > MN \cdot BO. \tag{6.14}$$

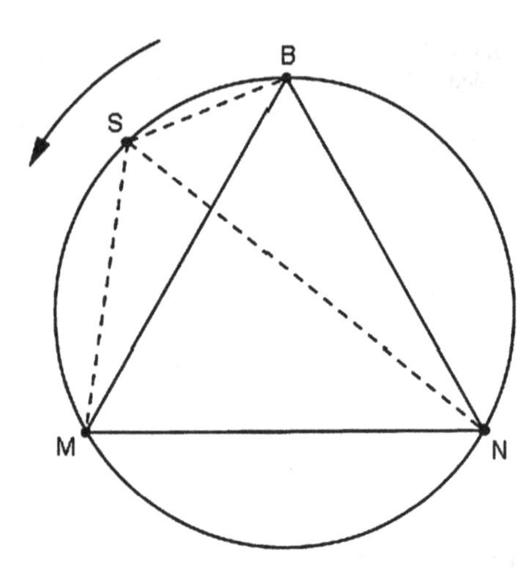

Fig. 6.2 Illustration of Problem 6.1.2

Since

$$BO > BB', \tag{6.15}$$

we have

$$N'M' > NM. \tag{6.16}$$

We have thus shown that $N'M'$ cannot be shorter than MN.

It remains to be shown that when the point M moves in the perimeter of the triangle ABC with orientation from A to B, and N is on the perimeter of ABC, the straight line segment MN covers all the interior points of the triangle ABC, as well as the points of the perimeter of the triangle ABC. This is the case since the positions of MN are in a one-to-one correspondence with the positions created if we keep MN constant and we let the point B to move on the constant arc from B to M (see Fig. 6.2), in a counterclockwise sense ($\widehat{B} = 60°$ and MN is a straight line segment of constant length).

Observing that

$$NS = BS + SM \geq BM = MN, \tag{6.17}$$

one can see that all the interior of the triangle ABC, including the perimeter, is covered. □

6.1.3 Let $PBCD$ be a rectangle inscribed in the circle (O, R). Let DP be an arc of (O, R) which does not contain the vertices of $PBCD$ and let A be a point of DP. The line parallel to DP that passes through A intersects the straight line BP at the point Z. Let F be the intersection point of the straight lines AB and DP and let Q be the intersection point of ZF and DC. Show that the straight line AQ is perpendicular to the straight line segment BD.

Solution The fact that we have to show perpendicularity leads us to think of how to use the orthocenter of a triangle. For this purpose, we will create a problem equiv-

Fig. 6.3 Illustration of Problem 6.1.3

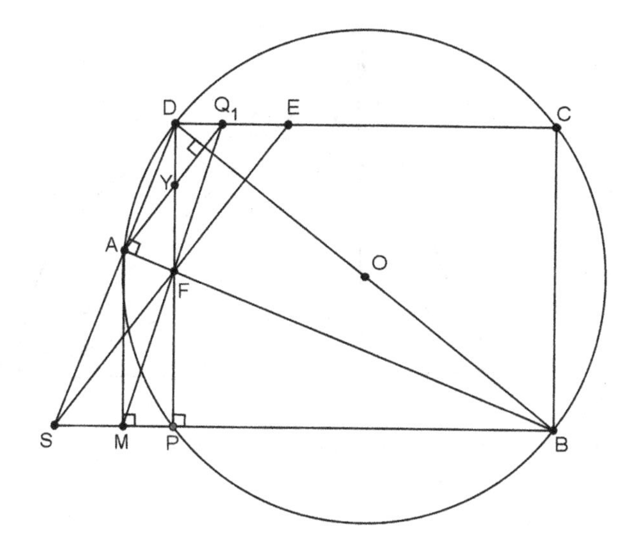

alent to the given one. In this problem, we will have to use the uniqueness of the position of a point or a few points when these satisfy certain conditions. We shall solve an equivalent problem on the same figure (see Fig. 6.3).

The problem Let $PBCD$ be a rectangle inscribed in the circle (O, R). Let DP be an arc of (O, R) which does not contain the vertices of $PBCD$ and let A be a point of DP. Consider the point Q_1 on DC such that $AQ_1 \perp DB$. If $M = Q_1F \cap BP$, show that the line $AM \parallel DP$.

If we show the above property, then the points M and Z will coincide and thus the points Q_1 and Q will coincide, which completes the proof.

Proof Let $S = EF \cap BP$ and let $Y = DF \cap AQ_1$. The third height of the triangle SBD lies on the line SE, and therefore

$$SF \perp DB,$$

and so

$$SE \parallel AQ_1.$$

Let $E = SF \cap DC$. Then the triangles FMS and FEQ_1 are similar, and therefore

$$\frac{MF}{FQ} = \frac{SF}{FE} = \frac{AY}{YQ_1}. \tag{6.18}$$

Hence

$$AM \parallel DP. \tag{6.19}$$

□

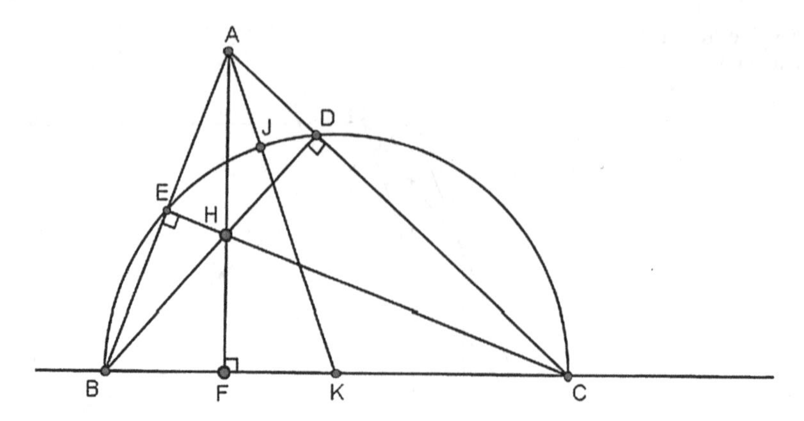

Fig. 6.4 Illustration of Problem 6.1.3 (Comment 2)

Remarks

1. In the above proof, we have used the fact that in Euclidean Geometry, only one straight line can be drawn through any point that does not belong to the given straight line, parallel to a given straight line in a plane. This led to the coincidence of the points M and Z.
2. An interesting problem that involves the use of orthocenters as well is the following:

Problem Let l be a straight line and A be a point not on l. Construct the perpendicular line from A to l by using the compass only once and the straightedge as many times needed.

Assume that the construction is achieved. Furthermore, suppose that a triangle ABC with points B and C on the line l and heights BD, CE is constructible, then the orthocenter could be determined. It suffices to connect it with the point A (see Fig. 6.4). This gives rise to the idea of the circle since the points B, C, D, E all lie on the circle diameter BC.

Construction We consider a point K on the straight line l and a point J in the interior of AK. We construct a semicircle with center K and radius KJ that lies on the half-plane containing A with respect to l. The semicircle intersects l at the points B and C. We find the intersections of AC, AB with the semicircle and denote them D and E, respectively. We then find the intersection of BD with CE which we denote by H. Finally, we connect the points A and H and we construct the perpendicular. □

6.1.4 Let l be a straight line and H be a point not lying on l. Let Ω be the set of triangles that have their orthocenter at H and let ABC be one of these triangles. Let l_1, l_2, l_3 be the reflections of the line l with respect to the sides BC, CA, AB. Let

$$A_1 = l_2 \cap l_3, \qquad B_1 = l_3 \cap l_1, \quad \text{and} \quad C_1 = l_1 \cap l_2.$$

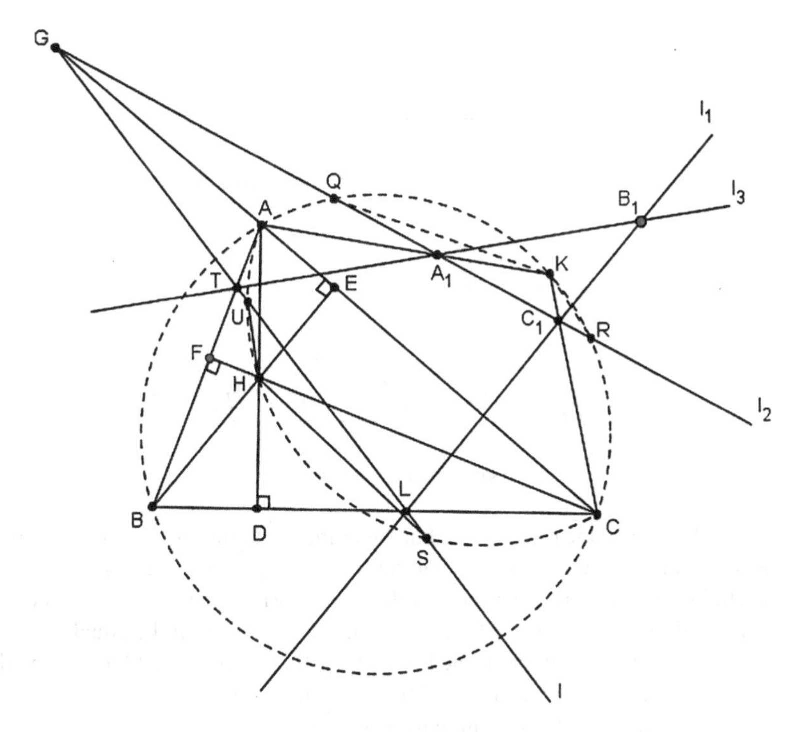

Fig. 6.5 Illustration of Problem 6.1.4

Prove that the ratio of the perimeter of the triangle $A_1B_1C_1$ to the area of the triangle $A_1B_1C_1$ is constant.

Solution We know that for any triangle, the ratio of its area to its semi-perimeter is equal to the radius of its inscribed circle (it is left as an exercise for the reader).

We shall therefore show that the circle inscribed in the triangle has a constant radius (see Fig. 6.5).

We observe that the points A, A_1, K and C, C_1, K are collinear when K is the intersection of the bisectors of the triangle $A_1B_1C_1$.

This is the case because A is the intersection of the bisectors of GTA, where T is the intersection of the lines AB, l, l_3.

In this case, K lies on the bisector AA_1 and C is the intersection of the bisector of $\widehat{C_1GL}$ with the bisector of the exterior angle $\widehat{C_1LG}$, where L is the intersection of the lines BC, l, l_1.

Therefore, the line CC_1 is the bisector of the exterior angle $\widehat{GC_1L}$ when G is the intersection of AC, l, l_2. This means that the point K belongs to the bisector CC_1.

We observe that

$$\widehat{A_1KC_1} = 90^\circ - \frac{\widehat{A_1B_1C_1}}{2}$$
$$= \frac{180^\circ - (180^\circ - \widehat{B_1TL} - \widehat{B_1LT})}{2}$$
$$= \frac{\widehat{B_1TL} + \widehat{B_1LT}}{2}.$$

Therefore,

$$\widehat{A_1KC_1} = \frac{180^\circ - 2\widehat{BTL} + 180^\circ - 2\widehat{TLB}}{2}$$
$$= 180^\circ - \widehat{ABC}. \tag{6.20}$$

Equality (6.20) leads to the conclusion that the point K belongs to the circumscribed circle of the triangle ABC. We know that the reflections of the orthocenter over the sides of ABC lie on the circumscribed circle as well. This means that H lies on an arc symmetrical to the arc AKC. These symmetrical arcs ought to be equal.

Let Q, R be the intersections of the line l_2 with the arc AKC and let U, S be the intersections of the line l with the arc AHC. The lines l, l_2 are reflections of each other over AC. Because of these symmetries, we have

$$QR = US. \tag{6.21}$$

We observe that

$$\widehat{KRQ} = \widehat{KC_1A_1} - \widehat{C_1KR}$$
$$= \frac{180^\circ - \widehat{GC_1L}}{2} - \frac{\text{arc } CR}{2}. \tag{6.22}$$

We also have

$$\widehat{KC_1A_1} = 90^\circ - \frac{\widehat{GC_1L}}{2} \tag{6.23}$$

and

$$\widehat{GCL} = \frac{180^\circ - \widehat{C_1LG}}{2} - \frac{\widehat{C_1GL}}{2} \tag{6.24}$$
$$= 90^\circ - \frac{\widehat{C_1LG} + \widehat{C_1GL}}{2}, \tag{6.25}$$

and hence

$$\widehat{KC_1A_1} + \widehat{GCL} = 180^\circ - 90^\circ = 90^\circ. \tag{6.26}$$

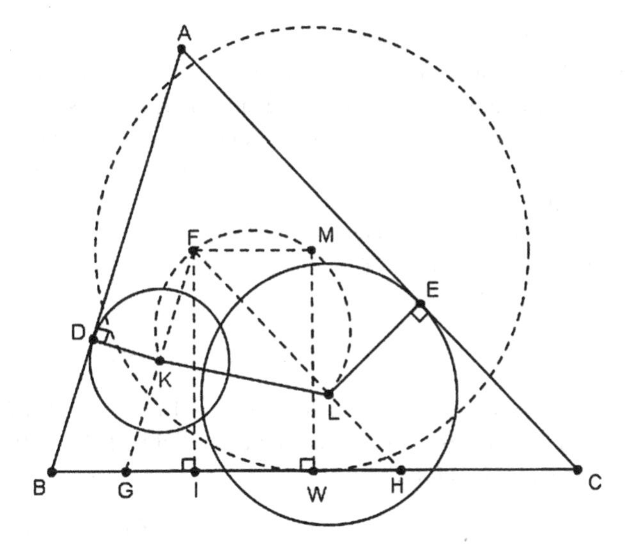

Fig. 6.6 Illustration of Problem 6.1.5

Therefore,

$$\widehat{KC_1A_1} = \widehat{HAC}. \tag{6.27}$$

By (6.22) and (6.27), we obtain arc $CR =$ arc SC. By the symmetry already mentioned, we have

$$\widehat{KRA_1} = \widehat{HUS}. \tag{6.28}$$

Similarly, we obtain

$$\widehat{RQK} = \widehat{USH}. \tag{6.29}$$

So, the triangles KQR and HSU are equal, and therefore their corresponding heights are equal. Since the distance of H from the line l remains constant, it follows that the radius of the circle inscribed in the triangle $A_1B_1C_1$ is constant since it coincides with the height of the triangle KA_1C_1 from the vertex K. □

6.1.5 Let ABC be a triangle. Consider two circles (K, R) and (L, r) with constant radii, which move in such a way that they remain tangent to the sides AB and AC, respectively, such that their centers belong to the interior of ABC, and finally such that the length of KL is preserved. Prove that there is a circle (M, h) (with constant radius) that moves in such a way that it remains tangent to the side BC and such that the triangle MKL has sides of constant length.

Solution In order to use the movement of the circles (K, R) and (L, r), we take into consideration the fact that their centers K, L, move in such a way that the distances R, r from the sides AB, AC, respectively, remain constant (see Fig. 6.6).

Let G be the intersection of BC with the parallel line to the side AB at distance R from AB and let H be the intersection of BC with the parallel line to the side AC at distance r from AC.

Let F be the intersection of these two parallel lines. We observe that the triangle FGH remains constant and is similar to the triangle ABC.

The points K, L move on the sides FG, FH, respectively, and KL has constant length. The angle $\widehat{KFL} = \widehat{BAC}$ is constant and since the arc KFL corresponds to the segment KL under a constant angle, the arc KFL is constant, too.

The fact that the arc KFL preserves constant length has an important consequence that the sides of the triangle MKL have constant length, where M is the intersection of the parallel line to BC with the arc KFL. This is the case because of the following reasoning. We have

$$\widehat{LFM} = \widehat{FHG} \tag{6.30}$$

since the angles are alternate interior. Also, the angles $\widehat{FHG}$ and $\widehat{ACB}$ are corresponding angles with respect to the parallel half-lines HF and CA as they intersect with CH. Therefore,

$$\widehat{FHG} = \widehat{ACB}. \tag{6.31}$$

We obtain

$$\widehat{MKL} = \widehat{LFM} = \widehat{ACB}, \tag{6.32}$$

and since the angle remains constant, it follows that the arc LM has constant length and thus the segment LM has constant length. Therefore,

$$\widehat{KML} = \widehat{KFL} = \widehat{BAC} \tag{6.33}$$

and

$$\widehat{MKL} = \widehat{MFL} = \widehat{ACB}. \tag{6.34}$$

Hence, the sides of the triangle MKL have constant length.

Considering a point W on BC such that $MW \perp BC$, we observe that the quadrilateral $FIWM$ is a rectangle. Therefore,

$$MW = FI. \tag{6.35}$$

Set $h = FI$, which is constant. The circle (M, h) satisfies the requirements of the problem. □

Method *For the proof of the equality of two planar shapes S, S', given the equalities*

$$P_1 = P_1', \qquad P_2 = P_2', \quad \ldots, \quad P_n = P_n',$$

where P_i, P_i' for $i = 1, 2, \ldots, n$ are elements of the shapes S and S', *respectively, it suffices to prove that when one can construct S by the use of $P_1, P_2, \ldots, P_n$, then S is uniquely defined.*

Example 6.1.1 Let ABC and $A'B'C'$ be two triangles. Assume that

$$BC = B'C', \qquad \widehat{A} = \widehat{A'},$$

and

$$\frac{AC}{AB} = \frac{A'C'}{A'B'}.$$

Prove that the triangles are equal.

Solution For the proof of the equality of the triangles, it is sufficient to show that if the triangle ABC can be constructed by the use of the elements

$$BC = a, \qquad \widehat{A} = \phi, \quad \text{and} \quad \frac{AC}{AB} = \frac{m}{n},$$

where a, m, n are given line segments and ϕ a given angle, then ABC is uniquely defined.

Let us assume that the triangle ABC has been constructed. We observe that for its vertex A we have:

1. It belongs to a constant arc C_1, which is the geometrical locus of the points E such that the angle $\widehat{BEC}$ is equal to the given angle ϕ, as well as to its symmetrical, with respect to the line BC, arc C_2.
2. It belongs to the circle C which is the geometrical locus of the points M such that

$$\frac{AC}{AB} = \frac{m}{n} \quad \text{(Apollonius circle)}.$$

The center of this circle belongs to the straight line defined by the line segment BC.

The circle C intersects the arcs C_1, C_2 in two points A, A'. Hence, we obtain two triangles ABC and $A'BC$, which are equal since they are symmetrical with respect to BC. Therefore, the triangle ABC is uniquely defined. □

6.1.6 Let ABC and $A_1B_1C_1$ be triangles. Let AD and A_1D_1 be bisectors of the angles $\widehat{A}$ and $\widehat{A_1}$, respectively, and let CE and C_1E_1 be the distances of the vertices C, C_1 from the lines AD and A_1D_1, respectively. Suppose that

$$AD = A_1D_1,$$
$$\widehat{CBA} = \widehat{C_1B_1A_1},$$
$$CE = C_1E_1.$$

Prove that

$$ABC = A_1B_1C_1.$$

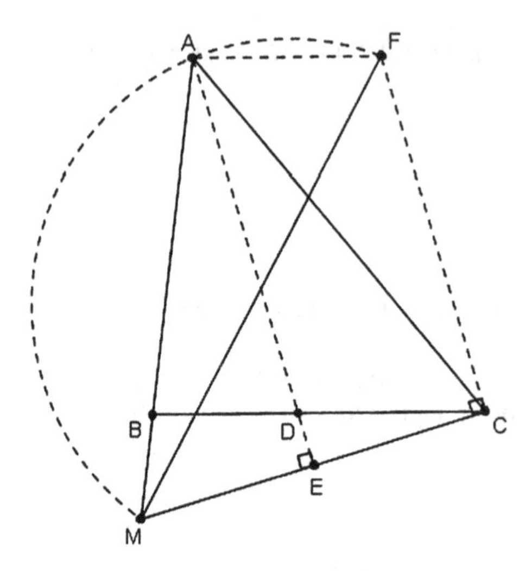

Fig. 6.7 Illustration of Problem 6.1.6

Solution We are going to prove an equivalent statement:

The construction of a triangle ABC is unique when the angle $\widehat{CBA}$, the length of the bisector AD, and the distance of the vertex C from the line containing the bisector AD are given.

Suppose that the triangle ABC has been constructed. We therefore have a triangle ABC with the given angle $\widehat{CBA}$, bisector AD, and distance CE of the vertex C from the line AD (see Fig. 6.7).

Let $M = AB \cap CE$. Then M is the reflection of C over AD. We consider the straight line segment CF such that

$$CF \perp CM \quad \text{and} \quad CF = AD.$$

The quadrilateral $ADCF$ is a parallelogram. The triangle FMC is constructible since it is a right triangle with its perpendicular sides MC and CF known. Therefore, the vertex A should lie on the perpendicular bisector of the segment MC since the triangle AMC is isosceles. But from the parallelogram $ADCF$ we have that

$$AE \parallel FC.$$

Therefore,

$$\widehat{MAF} = 180^\circ - \widehat{CBA}.$$

This implies that the arc MAF is constructible since it is the geometrical locus of the points that correspond to the constant straight line segment MF for the constant angle $180^\circ - \widehat{CBA}$.

The intersection of this arc with the perpendicular bisector of MC yields a unique point, the vertex A. The constructed triangle ABC is unique. □

6.1.7 In the triangle ABC let B_1, C_1 be the midpoints of the sides AC and AB, respectively, and H be the foot of the altitude passing through the vertex A. Prove

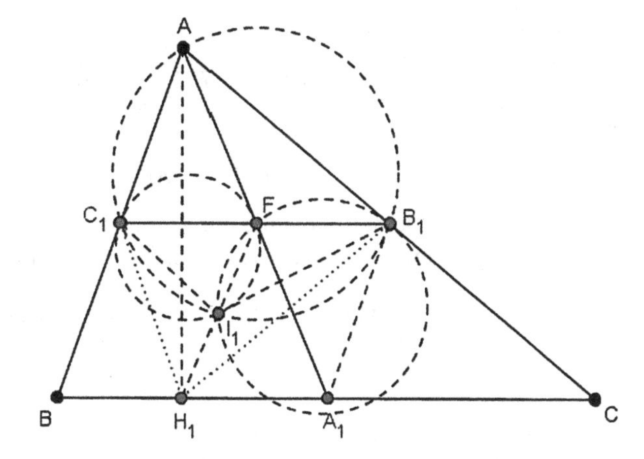

Fig. 6.8 Illustration of Problem 6.1.7

that the circumcircles of the triangles AB_1C_1, BC_1H, and B_1CH have a common point I and the line HI passes through the midpoint of the line segment B_1C_1.

(*Shortlist, 12th IMO, 1970, Budapest–Keszthely, Hungary*)

Solution We will use the following:

Lemma 6.1 *Given two triangles ABC and DEZ with $\widehat{BAM} = \widehat{EDN}$, $\widehat{MAC} = \widehat{NDZ}$, where M and N are the midpoints of the corresponding sides BC and EZ, the triangles ABC and DEZ are similar.*

(*The proof of the lemma is left as an exercise to the reader.*)

Let F be the midpoint of B_1C_1 (see Fig. 6.8). Consider the circle passing through the points C_1, F, tangent to the side AB and the circle determined by the points F, B_1, tangent to the side AC. Let I_1 be the second point of intersection of these two circles. This point I_1 exists since, if the circles had only one point in common, they would have a tangential contact at the point F. In this case, we consider the common tangent of the two circles at the point F, which intersects the sides AC and AB at the points S, T, respectively. By using the fact that the inscribed angle in a circle is equal to the angle which is formed by its corresponding chord and the tangent of the circle at the end of this chord, we have

$$\widehat{SB_1F} = \widehat{TC_1F},$$

and thus the line AC_1 should be parallel to the line AB_1. This is a contradiction.

The points A, F, A_1 are collinear when the point A_1 is the midpoint of the side BC. We observe that

$$\widehat{FI_1C_1} = \widehat{FC_1A} = \widehat{B}.$$

Therefore, the points B, H_1, I_1, C_1 belong to the same circumference (are *homocyclic*). Similarly, we deduce that the points C, B_1, I_1 and H_1 are homocyclic, where $H_1 \equiv FI_1 \cap BC$.

We have

$$\widehat{FI_1C_1} = \widehat{FC_1A} = \widehat{ABC}$$

and

$$\widehat{B_1I_1F} = \widehat{AB_1F},$$

where F is the midpoint of both the line segments C_1B_1 and AA_1. By the previous Lemma 6.1, it follows that the triangles $I_1B_1C_1$ and A_1B_1A are similar. Hence

$$\widehat{A_1H_1F} = \widehat{C_1FH_1} = \widehat{A_1FB_1} = \widehat{FA_1H_1} \tag{6.36}$$

and

$$C_1F \parallel H_1A_1. \tag{6.37}$$

Consequently, the triangle FH_1A_1 is isosceles, and thus $H_1F = FA_1 = FA$, that is, the triangle AH_1A_1 is orthogonal with $\widehat{A_1H_1A} = 90°$. Thus AH_1 is an altitude of the triangle ABC, which implies that the point H_1 coincides with the point H, and therefore the point I_1 must coincide with the point I because

$$\widehat{B_1I_1C_1} = \widehat{AB_1C_1} + \widehat{B_1C_1A} = 180° - \widehat{BAC}.$$

This completes the proof. □

6.1.8 Let ABC be an acute triangle and AM be its median. Consider the perpendicular bisector of the side AB and let E be its common point with the median AM. Let also D be the intersection of the median AM with the perpendicular bisector of the side AC. Suppose that the point L is the intersection of the straight lines BE and CD and that L_1, L_2 are the projections of L to AC and AB, respectively. Prove that the straight line L_1L_2 is perpendicular to AM.

Solution Since $BM = MC$ and AA' is a common altitude to both triangles ABM, AMC, it follows that (see Fig. 6.9)

$$S_{ABM} = S_{AMC} = \frac{S_{ABC}}{2}. \tag{6.38}$$

We know that when an angle of one triangle is equal to the angle of another triangle or of its supplementary angle, then the ratio of their areas is equal to the ratio of the product of their sides forming these angles. We have

$$\widehat{ACL} = \widehat{MAC} \quad \Rightarrow \quad \frac{S_{ALC}}{S_{AMC}} = \frac{LC \cdot AC}{AM \cdot AC} = \frac{LC}{AM} \tag{6.39}$$

and

$$\widehat{LBA} = \widehat{BAM} \quad \Rightarrow \quad \frac{S_{AMB}}{S_{ALB}} = \frac{AM \cdot AB}{AB \cdot BL} = \frac{AM}{BL}. \tag{6.40}$$

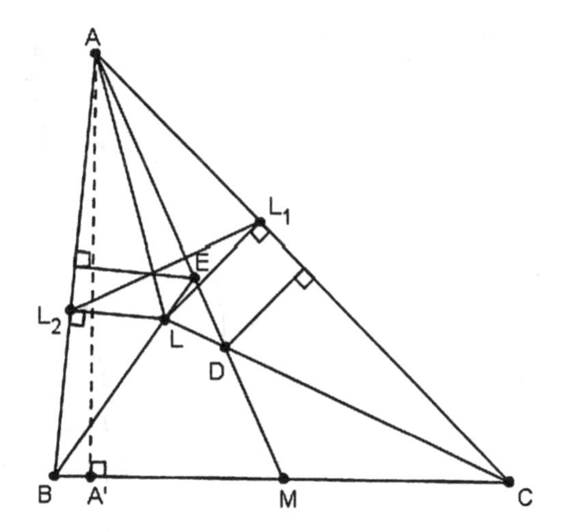

Fig. 6.9 Illustration of Problem 6.1.8

From (6.39), (6.40), and (6.38), we obtain

$$\frac{S_{ALC}}{S_{ALB}} = \frac{LC}{LB} = \frac{LC \cdot LA}{LB \cdot LA} \tag{6.41}$$

with

$$\begin{aligned}\widehat{ALB} + \widehat{CLA} &= 180^\circ - \widehat{LBA} - \widehat{BAL} + 180^\circ - \widehat{ACL} - \widehat{LAC} \\ &= 360^\circ - 2\widehat{A} \neq 180^\circ.\end{aligned} \tag{6.42}$$

Hence

$$\widehat{ALB} = \widehat{CLA}. \tag{6.43}$$

Since

$$\widehat{AL_1L} + \widehat{LL_2A} = 90^\circ + 90^\circ = 180^\circ$$

it follows that the quadrilateral AL_2LL_1 is inscribed in a circle, and thus

$$\widehat{L_1LA} = \widehat{L_1L_2A}. \tag{6.44}$$

We obtain

$$\widehat{L_1LA} + \widehat{CLL_1} = \widehat{CLA} = \widehat{ALB}.$$

Thus

$$2(\widehat{L_1LA} + \widehat{CLL_1}) = \widehat{CLA} + \widehat{ALB} = 360^\circ - \widehat{BLC}.$$

Therefore,

$$2(\widehat{L_1LA} + \widehat{CLL_1}) = 360^\circ - \big(180^\circ - (\widehat{B} - \widehat{LBA} + \widehat{C} - \widehat{LCA})\big).$$

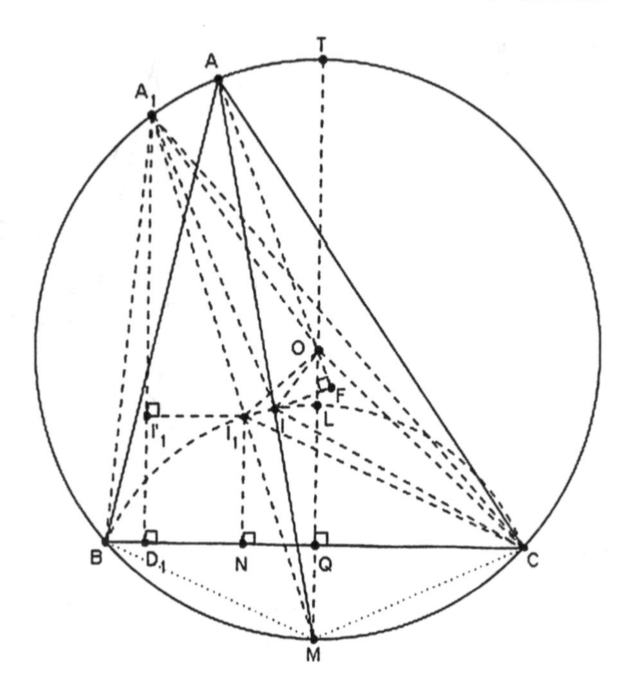

Fig. 6.10 Illustration of Problem 6.1.9

So

$$2(\widehat{L_1LA}+\widehat{CLL_1}) = 360^\circ - 2\widehat{A} \quad \Rightarrow \quad \widehat{L_1LA}+\widehat{CLL_1} = 180^\circ - \widehat{A}.$$

Thus

$$\widehat{L_1LA}+\widehat{BAE}+90^\circ-\widehat{LCA} = 180^\circ - (\widehat{EBA}+\widehat{LCA}).$$

Hence

$$\widehat{L_1L_2A}+\widehat{BAE} = 90^\circ, \tag{6.45}$$

and therefore the lines L_1L_2 and AM are perpendicular to each other. □

6.1.9 Prove that in a triangle with no angle larger than 90° the sum of the radii R, r of its circumscribed and inscribed circles, respectively, is less than the largest of its altitudes.

Solution Consider the circle (O, R) circumscribed around the triangle ABC. If the triangle is equilateral, then, because of the fact that

$$\widehat{A}=\widehat{B}=\widehat{C}=60^\circ,$$

we obtain

$$\max\{h_a, h_b, h_c\} = h_a = R + r, \tag{6.46}$$

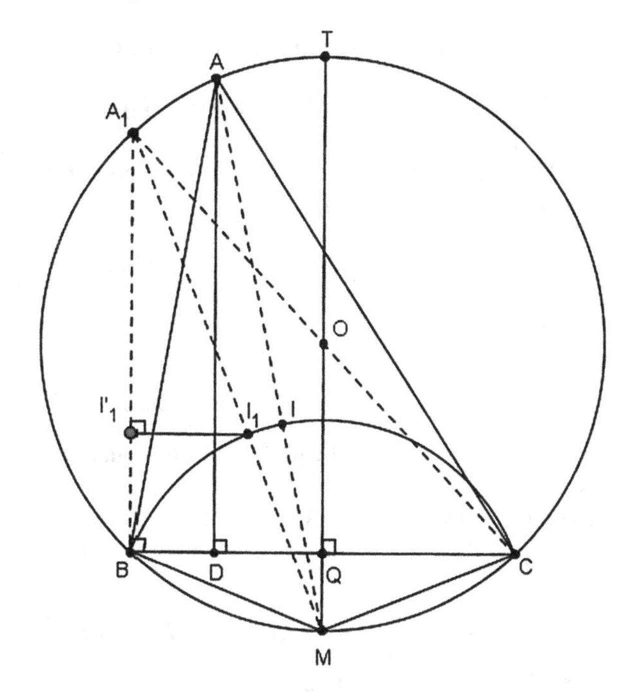

Fig. 6.11 Illustration of Problem 6.1.9 (Lemma 6.2)

since $R = 2r$ and $h_a = 3r$. In an equilateral triangle, the center of the circumscribed circle, the barycenter, and the orthocenter coincide (see Fig. 6.10). Suppose that

$$BC \leq AB \leq AC.$$

Then

$$\widehat{A} \leq \widehat{C} \leq \widehat{B}.$$

Thus

$$\widehat{A} \leq 60^\circ, \qquad \widehat{B} \geq 60^\circ, \quad \text{and} \quad h_a \geq h_c \geq h_b.$$

We shall make use of the following □

Lemma 6.2 *If a triangle ABC is inscribed in a circle* (O, R), *then the incenter I is identical to the common point of the bisector of the angle* $\widehat{A}$ *with the circle* (M, MB), *where* $MB = MC$, *and M is the midpoint of the arc BC which is seen by the angle* $\widehat{BAC}$ *(see Fig.* 6.11).

We observe that from the relation $\widehat{A} \leq 60^\circ$ one derives the inequalities $QM \leq QO$ and $QL \leq QO$, when L is the common point of the line OM with the circle (M, MB), when $Q = OM \cap BC$, since the condition $BC \perp OM$ holds true.

In the case $45^\circ \leq \widehat{A} \leq 60^\circ$, the isosceles triangle BA_1C $(BA_1 = BC)$ is the minimum non-obtuse triangle with I_1 as its incenter with

$$BC \leq AB \leq AC.$$

We note that the triangles A_1I_1O, COI_1 are equal. This implies that

$$\widehat{A_1OI_1} \geq 90^\circ$$

with I_1' being the foot of the projection of the point I onto the height A_1D_1.

It is a well known fact that

$$\widehat{D_1A_1I_1} = \widehat{MA_1O} = \frac{\widehat{CBA_1} - \widehat{A_1CB}}{2}.$$

Thus

$$A_1I_1' \geq OA_1 = R, \quad \text{and hence} \quad A_1D_1 \geq R + r_1,$$

given that $I_1'D_1 = r_1$, for the triangle A_1BC, where r_1 is the radius of its inscribed circle.

The non-obtuse triangle ABC has the property that its point A belongs to the arc TA_1 and the point M does not belong to this arc. This is a consequence of the relations

$$AB \leq BC = BA_1 \leq AC.$$

Then, if $h_a = AD$, the relations

$$h_a = AD \geq R + r$$

also hold, where D is the foot of the perpendicular from the point A onto the side BC.

Consequently, if $45^\circ \leq \widehat{A} \leq 60^\circ$, the inequality

$$\max\{h_a, h_b, h_c\} \geq R + r$$

holds true (see Fig. 6.11).

If $0 < \widehat{A} < 45^\circ$ then $A_1B = BC$ whenever $\widehat{CBA_1} = 90^\circ$. The point A of the non-obtuse triangle ABC, with $\widehat{A} \leq \widehat{C} \leq \widehat{B}$, is a point of the arc TA_1 with M not belonging to this arc. Thus $BC < BA_1$, and in this case, for the triangle A_1BC one gets

$$A_1I_1' = s - BC,$$

where s is the half of the perimeter of the triangle A_1BC.

We note that

$$A_1I_1' > R$$

is equivalent to

$$s - BC > R,$$

and this is equivalent to

$$\frac{A_1B + BC + 2R}{2} - BC > R,$$

which holds if and only if

$$A_1B > BC,$$

hence if F is the foot of the point I onto the height AD, we get

$$AI \geq A_1I_1,$$

$$\widehat{DAI} < \widehat{I_1'A_1I_1},$$

and so

$$AF \geq A_1I_1' > R.$$

This actually results in the given assumption in the case under consideration. Thus

$$h_a \geq A_1B > R + r.$$

Therefore, the proof is completed. □

6.1.10 Let KLM be an equilateral triangle. Prove that there exist infinitely many equilateral triangles ABC, circumscribed around the triangle KLM such that

$$K \in AB, \quad L \in BC \quad \text{and} \quad M \in AC \tag{6.47}$$

with

$$KB = LC = MA.$$

Solution Consider the center O of the triangle KLM and the circles (OKL), (OML), (OKM). From the point L we draw a straight line intersecting the circles (OKL), (OLM) at the points B, C. Linking the point C with M, we find on the other circle the point A and we see that the points A, K, and B are colinear (see Fig. 6.12).

Since we want

$$\widehat{A} = \widehat{B} = \widehat{C} = 60^\circ \quad \text{with } \widehat{K} = \widehat{L} = \widehat{M} = 60^\circ \tag{6.48}$$

to occur, we get the equality of the triangles:

$$AKM = KBL = MLC. \tag{6.49}$$

This is satisfied since

$$\widehat{AKM} + \widehat{LKB} = \widehat{BLK} + \widehat{LKB} = 120^\circ,$$

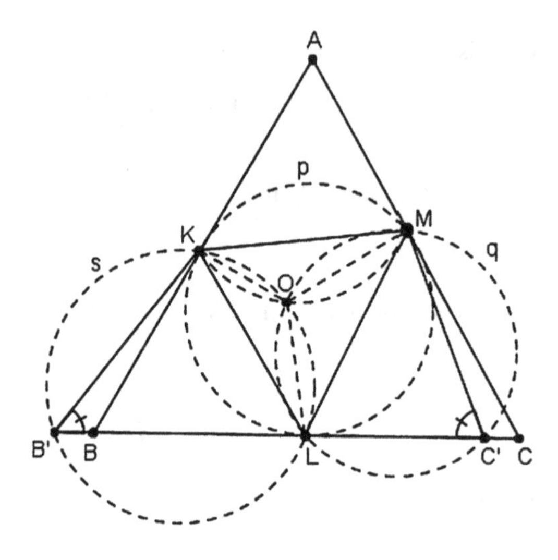

Fig. 6.12 Illustration of Problem 6.1.10

and so on. Furthermore, if O is the center of the prescribed circle of the triangle KLM, the circles

$$(KLM), \quad (KAM), \quad (KBL), \quad \text{and} \quad (LCM)$$

are equal. □

6.1.11 Let ABC be a triangle. Consider the points

$$K \in AB, \quad L \in BC, \quad M \in AC$$

such that

$$KB = LC = MA.$$

If the triangle KLM is equilateral, prove that the same holds true for the triangle ABC (see Fig. 6.13).

Solution Suppose that the triangle ABC is not equilateral. Then at least one of its angles is greater than or equal to 60° and at least another one of them shall be less than or equal to 60°. Let $\widehat{B} > 60°$ and $\widehat{C} < 60°$. Consider the center O of the equilateral triangle KLM. Observe that

$$\widehat{KOL} = \widehat{LOM} = \widehat{MOK} = 120°.$$

Consider also the equal circles (KML), (OKM), (OKL), (OLM). Indeed, since

$$\widehat{KOL} = \widehat{LOM} = \widehat{MOK} = 120°$$

and

$$KL = MK = LM,$$

and the triangle KLM is equilateral, the equality of these circles follows.

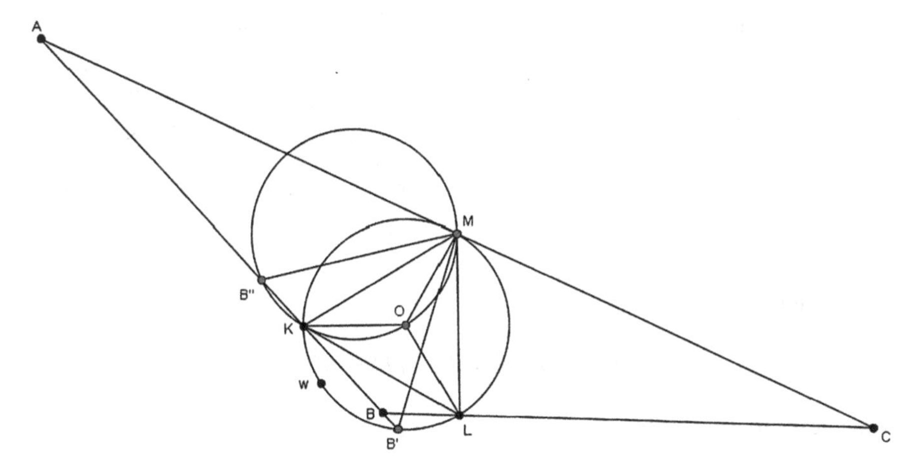

Fig. 6.13 Illustration of Problem 6.1.11

According to what we have already mentioned, the circle (OKL) should meet the straight semiline LB at a point B' (see Fig. 6.12) such that

$$\widehat{KB'L} = 180^\circ - \widehat{KOL} = 180^\circ - 120^\circ = 60^\circ$$

with $\widehat{B} > 60^\circ$, and thus $LB' > LB$. Similarly, the circle (MOL) should intersect the straight semiline LC at a point C' such that $LC' < LC$. We observe that

$$\widehat{B'KL} = 120^\circ - \widehat{KLB'} = \widehat{C'LM}$$

and, of course,

$$\widehat{LB'K} = \widehat{MC'L} = 60^\circ$$

with $KL = LM$. It follows that the triangles $KB'L$ and MLC' are equal, and hence $KB' = LC'$. Simultaneously, we have

$$LC' < LC = BK < KB',$$

and the contradiction is evident (in case $\widehat{B} < 120^\circ$).

Suppose now that $\widehat{B} \geq 120^\circ$. In this case, we see that KL, LM, MK are tangent to the circles they *contact*. The point B shall belong either to the minimal circular segment KWL of chord KL of the circumscribed circle to the triangle KLM, or be inside the circular segment $KWLK$.

Let B', B'' be the intersections of the straight line AB with the circles (KLM) and (OKM), respectively. The triangle $MB'B''$ is equilateral since

$$\widehat{B'} = \widehat{B''} = 60^\circ,$$

and hence

$$KB < B'B'' = MB''$$

with $\widehat{MB''A} = 120^\circ > 90^\circ$ since $\widehat{A} < 60^\circ$.

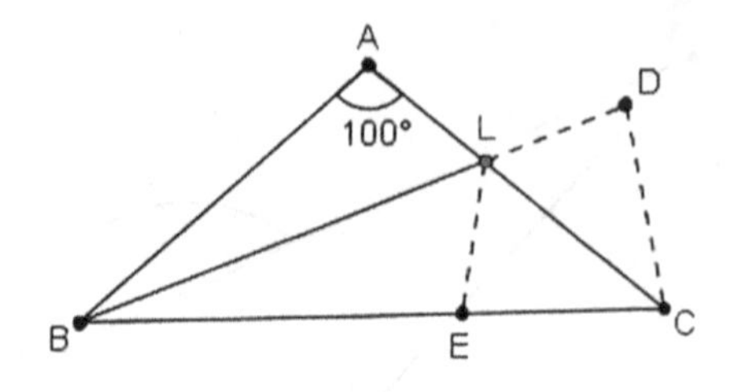

Fig. 6.14 Illustration of Problem 6.1.12

Therefore, we get

$$KB < B'B'' = MB'' < MA = KB,$$

which leads to a contradiction. □

6.1.12 Let ABC be an isosceles triangle with $\widehat{A} = 100°$. Let BL be the bisector of the angle $\widehat{ABC}$. Prove that

$$AL + BL = BC.$$

(Proposed by Andrei Razvan Baleanu [23], Romania)

Solution (by Miguel Amengual Covas, Cala Figuera, Mallorca, Spain) Let D be a point on BL beyond L such that $LD = LA$ and let E be a point on BC such that LE bisects the angle $\widehat{BLC}$.

Because of the fact that $\widehat{ABC} = \widehat{BCA} = 40°$, we obtain (see Fig. 6.14)

$$\widehat{ABL} = \widehat{LBE} = 20°, \qquad \widehat{BLA} = 60°, \qquad \widehat{BLC} = 120°,$$

and

$$\widehat{BLE} = \widehat{ELC} = \frac{1}{2}\widehat{BLC} = 60° = \widehat{DLC}.$$

Thus the triangles ABL and EBL are congruent (angle-side-angle), which implies $LA = LE$. Therefore, $LD = LE$. We also have that LC is the bisector of the angle $\widehat{ELD}$ in the isosceles triangle DLE. Hence LC is the perpendicular bisector of the base DE. Therefore,

$$\widehat{LCD} = \widehat{ECL} = 40°$$

and

$$\widehat{EDC} = 90° - \widehat{LCD} = 50°.$$

Hence

$$\widehat{BDC} = \widehat{BDE} + \widehat{EDC} = 30° + 50° = 40° + 40° = \widehat{BCL} + \widehat{LCD} = \widehat{BCD}.$$

Fig. 6.15 Illustration of Problem 6.1.13

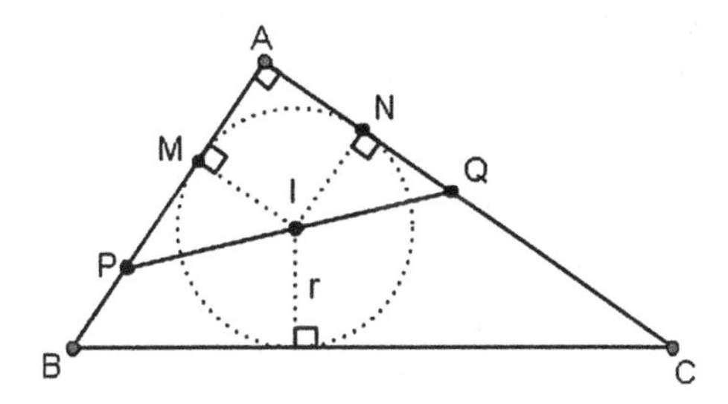

Thus BCD is an isosceles triangle with the property

$$BC = BD = BL + LD = BL + LA.$$

This completes the proof. □

6.1.13 Let ABC be a right triangle with $\widehat{A} = 90°$ and d be a straight line passing through the incenter of the triangle and intersecting the sides AB and AC at the points P and Q, respectively. Find the minimum of the quantity $AP \cdot AQ$.

(*Proposed by Dorin Andrica [17], Romania*)

Solution (by Athanassios Magkos, Greece) Let I be the incenter of the triangle ABC. Assume that M, N are the projections of I on AB and AC, respectively (see Fig. 6.15). We have $IM = IN = r$. From the similarity of the triangles PMI, INQ, we obtain

$$PM \cdot NQ = r^2. \tag{6.50}$$

By r we denote the inradius of the triangle ABC.

It follows that

$$\begin{aligned} AP \cdot AQ &= (AM + MP)(AN + NQ) \\ &= AM \cdot AN + AM \cdot NQ + MP \cdot AN + MP \cdot NQ \\ &= 2r^2 + r(NQ + MP) \geq 2r^2 + 2r\sqrt{MP \cdot NQ} \\ &= 2r^2 + 2r^2 = 4r^2. \end{aligned} \tag{6.51}$$

Thus

$$AP \cdot AQ \geq 4r^2. \tag{6.52}$$

The above becomes an equality if and only if

$$MP = NQ \quad \Leftrightarrow \quad AP = AQ \quad \Leftrightarrow \quad \widehat{APQ} = \widehat{AQP} = 45°.$$ □

6.1.14 Let P be a point in the interior of a circle. Two variable perpendicular lines through P intersect the circle at the points A and B. Find the geometrical locus of the midpoint of the line segment AB.

(*Proposed by Dorin Andrica [16], Romania*)

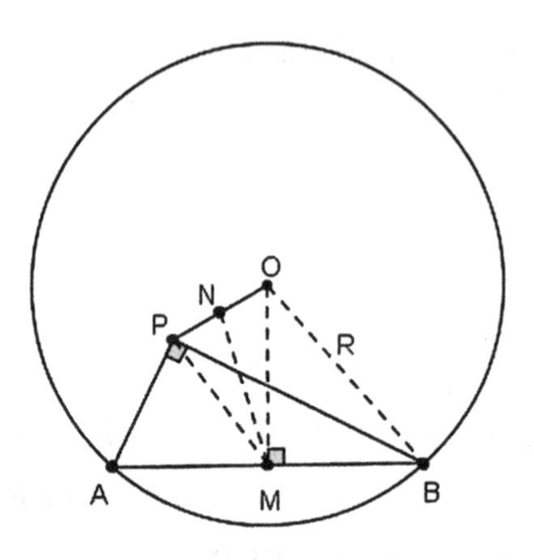

Fig. 6.16 Illustration of Problem 6.1.14

First solution (by G.R.A Problem Solving Group, Roma, Italy) Without loss of generality, we can assume that

$$P = t \in [0, 1]$$

and consider the circle

$$C = \{z \in \mathbb{C} : |z| = 1\}.$$

Let

$$A = z = x + iy \in \mathbb{C}.$$

Then

$$B = w = si(z - P) + P \in \mathbb{C} \quad \text{with some } s > 0.$$

Hence

$$1 = |z|^2 = (t - sy)^2 + s^2(x - t)^2. \tag{6.53}$$

The midpoint of the straight line segment AB is determined by (see Fig. 6.16)

$$M = \frac{A + B}{2}.$$

We claim that

$$\left|M - \frac{P}{2}\right| = \frac{\sqrt{2 - |P|^2}}{2}. \tag{6.54}$$

By (6.53), we have

$$\begin{aligned}\left(2\left|M - \frac{P}{2}\right|\right)^2 &= (x - sy)^2 + \left(s(x - t) + y\right)^2 \\ &= x^2 + y^2 + 1 - t^2 = 2 - t^2.\end{aligned} \tag{6.55}$$

Therefore, the required geometrical locus is a circle with center $\frac{P}{2}$ and radius

$$\frac{\sqrt{2-|P|^2}}{2}. \qquad \square$$

Second solution We could use the fact that the median of a right triangle passing from its vertex corresponding to the right angle equals the half of its hypotenuse. Since M is the midpoint of the chord AB, it follows that the straight line OM is perpendicular to this chord. Hence,

$$OM^2 + MB^2 = R^2,$$

so

$$OM^2 + MP^2 = R^2,$$

and thus

$$2MN^2 + \frac{OP^2}{2} = R^2.$$

Therefore

$$MN = \frac{\sqrt{2R^2 - OP^2}}{2}.$$

Hence we derive that the point M belongs to a circle with center at the point N and radius

$$\frac{\sqrt{2R^2 - OP^2}}{2}. \qquad \square$$

6.1.15 Prove that any convex quadrilateral can be dissected into n, $n \geq 6$, cyclic quadrilaterals.

(*Proposed by Dorin Andrica [19], Romania*)

Solution (by Daniel Lasaosa, Spain) Any convex quadrilateral is dissected into two triangles by either of its diagonals; any concave quadrilateral is dissected into two triangles by exactly one of its diagonals; any crossed quadrilateral is already formed by two triangles joined at one vertex, and where two of the sides of each triangle are on the straight line containing two of the sides of the other.

In the triangle ABC, let I be the incenter and D, E, F the points where the incircle touches the sides BC, CA, and AB, respectively. The triangle ABC may be dissected into three cyclic quadrilaterals $AEIF$, $BFID$, $CDIE$.

With no loss of generality, assume that the angle $\widehat{C}$ of the triangle ABC is acute. Consider the circumcenter O of the triangle ABC and take a point O' on the perpendicular bisector of AB that is closer to AB than O. The circle with center O' through A, B leaves C outside. Therefore, it must intersect the interior of the segments AC, BC at the points E, D or the quadrilateral $ABDE$ is cyclic.

Let us write $n = 3 + 3u + v$, where $u \geq 1$ is an integer and $v \in \{0, 1, 2\}$. Dissect any quadrilateral $ABCD$ in two triangles and in the following dissect one of them into three cyclic quadrilaterals.

If $v \neq 0$, dissect the other triangle into one cyclic quadrilateral and one triangle. If $v = 2$, dissect again this latter triangle into one cyclic quadrilateral and one triangle. After having performed this procedure, we have dissected the original quadrilateral into $3 + v$ cyclic quadrilaterals ($3, 4, 5$ for $v = 0, 1, 2$, respectively) and one triangle.

Dissect now this triangle into u triangles (for example, by dividing one of its sides in u equal parts and joining each point of division with the opposite vertex), and dissect now each one of these u triangles into three cyclic quadrilaterals.

We have thus dissected the original quadrilateral into $3 + v + 3u = n$ cyclic quadrilaterals. □

6.1.16 Let ABC be a triangle such that $\widehat{ABC} > \widehat{ACB}$ and let P be an exterior point in its plane such that

$$\frac{PB}{PC} = \frac{AB}{AC}. \tag{6.56}$$

Prove that

$$\widehat{ACB} + \widehat{APB} + \widehat{APC} = \widehat{ABC}. \tag{6.57}$$

(Proposed by Mircea Becheanu [25], Romania)

Solution (by Daniel Lasaosa, Spain) Note that the relation (6.56) defines an Apollonius circle γ with center on the line BC which passes through A and through the point D, where the internal bisector of the angle $\widehat{A}$ intersects BC, leaving the point B inside γ and the point C outside γ, since $\widehat{ABC} > \widehat{ACB}$.

The powers of the points B, C with respect to the circle γ are p_B, p_C, respectively (see Fig. 6.17). In addition,

$$\frac{p_B}{p_C} = \frac{BD}{CD} \cdot \frac{BD'}{CD'} = \frac{BA^2}{CA^2} = \frac{c^2}{b^2}, \tag{6.58}$$

where D' is the point diametrically opposite to D in γ. Assume that T, U are the second points where PB, PC meet γ (the first one being clearly P in both cases).

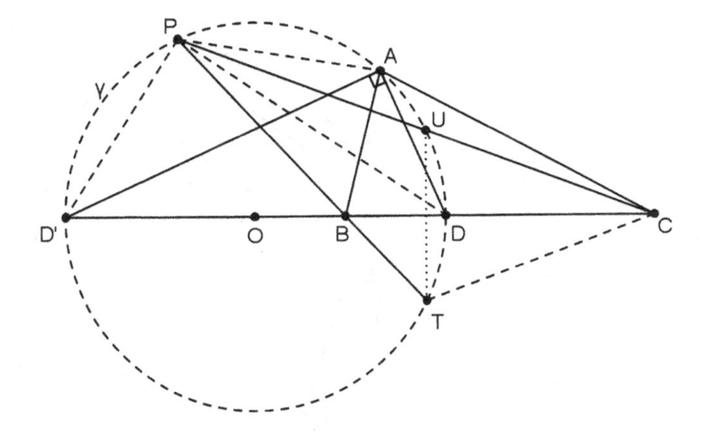

Fig. 6.17 Illustration of Problem 6.1.16

Therefore,

$$\begin{aligned}\frac{CT}{CU} &= \frac{b \cdot BT}{c \cdot CU} \\ &= \frac{b \cdot p_B}{c \cdot CU \cdot PB} \\ &= \frac{b^2 \cdot p_B}{c^2 \cdot CU \cdot PC} \\ &= \frac{b^2 \cdot p_B}{c^2 \cdot p_C} = 1, \end{aligned} \tag{6.59}$$

or

$$CT = CU,$$

and similarly

$$BT = BU.$$

Thus BC is the perpendicular bisector of TU, which is therefore symmetric with respect to BC. Therefore, if P is on the same half plane with A, with respect to the straight line BC, then

$$\begin{aligned}\widehat{APB} = \widehat{APT} &= 180° - \widehat{ADT} \\ &= 180° - \widehat{ADB} - \widehat{BDT} \\ &= 180° - \widehat{ADB} - \widehat{BDU} \\ &= 180° - 2\widehat{ADB} - \widehat{ADU} \\ &= 180° - 2\widehat{ADB} - \widehat{APU} \\ &= 180° - 2\widehat{ADB} - \widehat{APC}. \end{aligned} \tag{6.60}$$

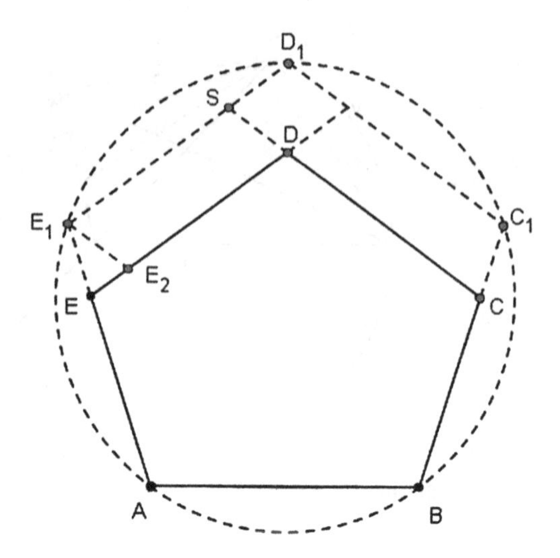

Fig. 6.18 Illustration of Problem 6.1.17

Similarly, we obtain the same result if P is on the opposite half plane. In either case, we have

$$\begin{aligned}\widehat{APB}+\widehat{APC} &= 180^\circ - 2\widehat{ADB} \\ &= 180^\circ - 2\left(180^\circ - \widehat{B} - \frac{\widehat{A}}{2}\right) \\ &= 2\widehat{B} + \widehat{A} - 180^\circ \\ &= \widehat{B} - \widehat{C} \qquad (6.61) \\ &= \widehat{ABC} - \widehat{ACB}, \qquad (6.62)\end{aligned}$$

hence

$$\widehat{ACB}+\widehat{APB}+\widehat{APC}=\widehat{ABC}.$$

This completes the proof. □

6.1.17 Prove that if a convex pentagon satisfies the following properties:

1. All its internal angles are equal;
2. The lengths of its sides are rational numbers,

then this is a regular pentagon.
(*18th BMO, 2001, Belgrade, Serbia*)

Solution The following facts are going to be used (see Fig. 6.18):

- The number $\sin 18^\circ$ is irrational.
- A convex pentagon with equal internal angles and more than two sides equal is a regular pentagon.

- An isosceles triangle ABC ($AB = AC$) with $\widehat{A} = 36°$ cannot have all its sides with lengths rational numbers since this contradicts the fact that $\sin 18°$ is an irrational number.

Let $ABCDE$ be a convex pentagon with

$$\widehat{A} = \widehat{B} = \widehat{C} = \widehat{D} = \widehat{E} = 108°$$

and with the lengths of its sides given by certain rational numbers. With no loss of generality, let us assume that

$$AB \geq BC, \qquad AB > CD, \qquad AB > DE, \qquad AB > EA.$$

Consider the regular pentagon $ABC_1D_1E_1$ (it might be $C \equiv C_1$). Let $E_1E_2 \parallel C_1D_1$. Hence, if E does not coincide with E_1, it follows that

(i) $\widehat{E} = \widehat{E_1} = 108°$.
(ii) The sides of the isosceles triangle E_1EE_2 are rational numbers since

$$E_1E_2 = E_1E = AB - AE$$

and EE_2 is rational (being the difference of two rational numbers).

This is actually a contradiction because of the observation–assumption we have made (clearly, $\widehat{EE_1E_2} = 108°$). Hence,

$$E = E_1 \quad \Rightarrow \quad D = D_1$$

and

$$C = C_1.$$

In conclusion, the pentagon $ABCDE$ happens to be a regular one, and this completes the proof. □

The following lemma deals with the irrationality of the trigonometric number $\sin 18°$, a fact that we have already used in the preceding problem.

Lemma 6.3 *The number* $\sin 18°$ *is irrational.*

Proof It holds

$$\sin(3 \cdot 18°) = \cos(2 \cdot 18°),$$

and therefore,

$$4\sin^3 18° - 2\sin^2 18° - 3\sin 18° + 1 = 0. \tag{6.63}$$

Let $x = \sin 18°$, then (6.63) assumes the form

$$4x^3 - 2x^2 - 3x + 1 = 0, \tag{6.64}$$

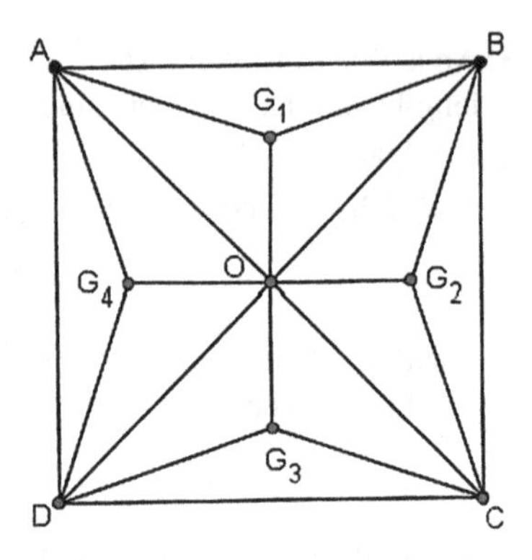

Fig. 6.19 Illustration of Problem 6.1.18

and the problem reduces to the study of the existence of a rational solution of (6.64). Suppose there exist $k, l \in \mathbb{Z}$, $l \neq 0$, $(k, l) = 1$ such that $x = k/l$ satisfies Eq. (6.64). Hence

$$\begin{aligned} 4k^3 &= \left(2k^3 + 3kl + l^2\right)l, \\ l^3 &= \left(3l^2 + 2kl - 4k^2\right)k. \end{aligned} \tag{6.65}$$

Using the first relation in (6.65), we deduce

$$l|4, \tag{6.66}$$

and thus $l \in \pm\{1, 2, 4\}$. From the second relation, we get

$$k = \pm 1. \tag{6.67}$$

Consequently, $x \in \pm\{1, \frac{1}{2}, \frac{1}{4}\}$ which are easily rejected as solutions of (6.64). This completes the proof of the assertion on the irrationality of $\sin 18°$. □

Since we investigated a matter of irrationality of a trigonometric function, it is useful to state a more general theorem.

Theorem 6.1 *The trigonometric functions are irrational at non-zero rational values of the arguments.*

(*Cf. I. Niven, Irrational Numbers, The Mathematical Association of America, Washington, D.C.*, 1956.)

6.1.18 Let k points be in the interior of a square of side equal to 1. We triangulate it with vertices these k points and the square vertices. If the area of each triangle is at most $\frac{1}{12}$, prove that $k \geq 5$.

(*Proposed by George A. Tsintsifas, Greece*)

Solution (by George A. Tsintsifas) Let p be the number of triangles of the triangulation of the unit square (see Fig. 6.19). The sum of the angles of the p triangles is equal to

$$4 \cdot 90° + k \cdot 4 \cdot 90°,$$

that is,

$$4 \cdot 90° + k \cdot 4 \cdot 90° = 2p \cdot 90°.$$

Thus $p = 2 + 2k$. Now, we have

$$E_1 + E_2 + \cdots + E_p = 1, \tag{6.68}$$

where E_i is the area of the ith triangle of the triangulation. According to (6.68), we get

$$E_1 + E_2 + \cdots + E_{2k+2} \leq (2k+2) \cdot \frac{1}{12}. \tag{6.69}$$

By (6.68), (6.69), we finally obtain

$$1 \leq \frac{2k+2}{12}, \tag{6.70}$$

or

$$5 \leq k. \qquad \square$$

Remark 6.1 A *triangulation of an n-gon* with the points on the perimeter (except the vertices) and k-internal points is the division of the polygon into triangles with vertices the $n + m + k$ points.

Remark 6.2 An example which refers to the equality with respect to the inequality obtained in Problem 6.1.18 is the following. Consider the square $ABCD$ and choose its center O together with the barycenters G_1, G_2, G_3, G_4 of the triangles AOB, BOC, COD, and DOA, respectively.

6.1.19 Let ABC be an equilateral triangle and D, E, F be points of the sides BC, CA, and AB, respectively. If the center of the inscribed circle of the triangle DEF is the center of the triangle ABC, determine what kind of triangle DEF is.

(*Proposed by George A. Tsintsifas, Greece*)

Solution (by George A. Tsintsifas) We shall prove that the triangle DEF is equilateral using the method of proof by contradiction. Let us assume that the triangle DEF is not equilateral (see Fig. 6.20). Then, two possibilities may occur:

(a) $\hat{D} \geq 60°$, $\hat{E}, \hat{F} \leq 60°$;
(b) $\hat{D} \leq 60°$, $\hat{E}, \hat{F} \geq 60°$.

First we examine Case (a). On the side EF and internally to the line segment EF, we can find the points E', F' such that if we draw the tangents to the inscribed circle of DEF then

$$\hat{E}' = \hat{F}' = 60°.$$

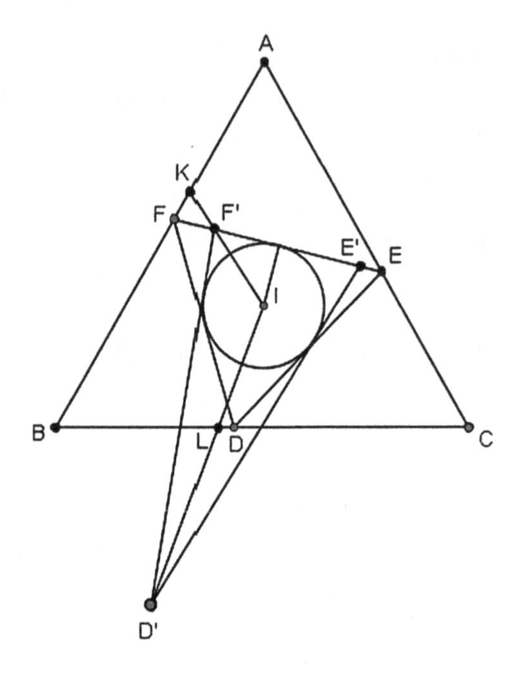

Fig. 6.20 Illustration of Problem 6.1.19

Let D' be the point of intersection of these two tangents. This point is outside the triangle ABC. Let

$$IF' \cap AB \equiv K, \qquad ID' \cap BC \equiv L.$$

It is easy to see that

$$IK = IL,$$
$$IF' < IK,$$
$$ID' > IL,$$

and thus we get a contradiction. Similarly, we exclude Case (b), and this completes the proof. □

6.2 Geometric Problems with More Advanced Theory

6.2.1 Consider a circle $C(K,r)$, a point A on the circle, and a point P outside the circle. A variable line l passes through the point P and intersects the circle at the points B and C. Let H be the orthocenter of the triangle ABC. Prove that there exists a unique point T in the plane of the circle $C(K,r)$ such that the sum

$$HA^2 + HT^2$$

remains constant (independent of the position of the line l).

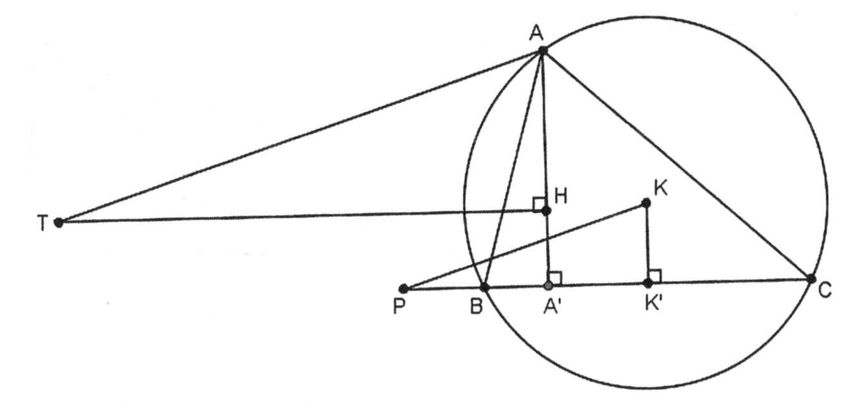

Fig. 6.21 Existence: Problem 6.2.1

Solution In the following, we will study the existence as well as the uniqueness of a point T such that the given hypothesis is satisfied (see Fig. 6.21).

Existence If $KK' \perp BC$, then it is known that

$$AH = 2KK'. \tag{6.71}$$

Bearing in mind that for the creation of the sum $HA^2 + HT^2$ it is enough to construct a triangle HAT with

$$\widehat{H} = 90°, \tag{6.72}$$

we try to construct a triangle HAT similar to $KK'P$. Thus from the point A we draw

$$AT \parallel KP \tag{6.73}$$

and such that $AT = 2KP$.
Then the triangles AHT and $KK'P$ are similar because

$$\widehat{K'KP} = \widehat{HAT} \tag{6.74}$$

and

$$\frac{AH}{KK'} = \frac{AT}{KP} = 2. \tag{6.75}$$

Hence,

$$\widehat{AHT} = \widehat{KK'P} = 90°, \tag{6.76}$$

and therefore

$$HA^2 + HT^2 = AT^2 = (2 \cdot KP)^2 = 4KP^2, \tag{6.77}$$

which is constant and independent of the position of the line l.

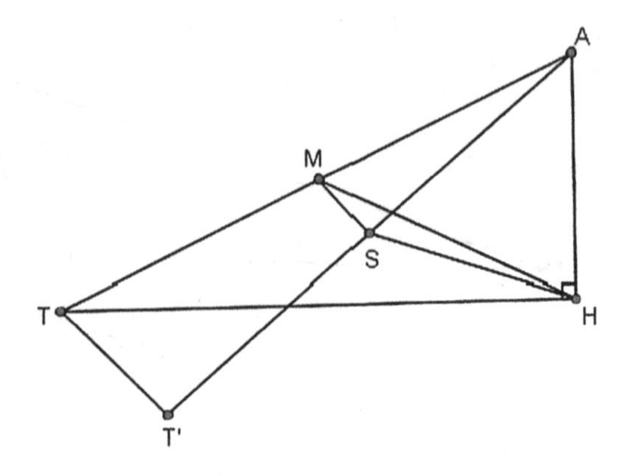

Fig. 6.22 Uniqueness: Problem 6.2.1

Uniqueness Suppose there exists a point $T' \neq T$ such that (see Fig. 6.22)

$$HA^2 + HT'^2 = c^2,$$

where c is constant, for every position of H (possibly $c^2 \neq KP^2$). Then the midpoints M, S of the segments AT, AT' respectively define a segment MS of constant length. Also, the lengths of the segments MH and SH are constant since, by the first theorem of medians applied to the triangle AHT', we have

$$MH = \frac{AT}{2} = KP \tag{6.78}$$

and

$$2HS^2 = c^2 = \frac{AT'^2}{2}. \tag{6.79}$$

Then the triangle MSH can be constructed with only two possible positions for the vertex H, which contradicts the fact that H can take an infinite number of positions depending upon the position of the line l. □

6.2.2 Consider two triangles ABC and $A_1B_1C_1$ such that

1. The lengths of the sides of the triangle ABC are positive consecutive integers and the same property holds for the sides of the triangle $A_1B_1C_1$.
2. The triangle ABC has an angle that is twice the measure of one of its other angles and the same property holds for the triangle $A_1B_1C_1$.

Compare the areas of the triangles ABC and $A_1B_1C_1$.

Solution Let the triangle ABC have sides

$$AC = b,$$

$$AB = c = b + 1,$$

$$BC = a = b - 1,$$

where b is a positive integer. We observe that

$$b - 1 > 0, \tag{6.80}$$

which is equivalent to

$$b > 1, \tag{6.81}$$

and that

$$\widehat{CAB} < \widehat{ABC} < \widehat{BCA}. \tag{6.82}$$

Let x be the length of the projection of the side AC onto AB, let y be the length of the projection of the side BC onto CA, and z be the length of the projection of AB onto BC. By using the standard formulas

$$h_k = \frac{2}{k}\sqrt{s(s-a)(s-b)(s-c)},$$

where $k \in \{a, b, c\}$, for the computation of the lengths of the heights of a triangle ABC with $BC = a$, $CA = b$, and $AB = c$, we obtain

$$\sqrt{s(s-a)(s-b)(s-c)} = \frac{b}{4}\sqrt{3\left(b^2-4\right)},$$

hence

$$x^2 = b^2 - \frac{3b^2(b^2-4)}{4(b+1)^2},$$

and thus

$$x = \frac{b}{2} + \frac{3b}{2b+1}. \tag{6.83}$$

Also,

$$y^2 = (b-1)^2 - \frac{3(b^2-4)}{4},$$

and thus

$$y = \frac{|b-4|}{2}. \tag{6.84}$$

Finally,

$$z^2 = (b+1)^2 - \frac{3b^2(b^2-4)}{4(b-1)^2},$$

and thus

$$z = \frac{b^2 + 2}{2(b-1)}, \tag{6.85}$$

where $AC = b$, $AB = b + 1$, and $BC = b - 1$.

Therefore, the numbers x, y, z are rational. Now, the quantity x/b is decreasing with respect to b. Therefore, when b increases, the angle $\widehat{CAB}$ increases, as well. From relation (6.82) we infer that

$$\widehat{CAB} \leq 45^\circ \tag{6.86}$$

because

$$\widehat{CBA} > 45^\circ.$$

Thus

$$\frac{1}{2} + \frac{3}{2(b+1)} < \frac{\sqrt{2}}{2},$$

that is,

$$b > 3\sqrt{2} + 2.$$

Then we have

$$\frac{|z|}{b-1} < \frac{\sqrt{2}}{2}, \tag{6.87}$$

and thus

$$\widehat{ABC} > 45^\circ. \tag{6.88}$$

Hence

$$\widehat{BCA} < 90^\circ. \tag{6.89}$$

This case is therefore rejected since it does not allow one angle to be twice as large as another. Therefore, $b \leq 6$, and thus $b \in \{2, 3, 4, 5, 6\}$. Suppose that one of the following cases holds true:

$$\widehat{BCA} = 2\widehat{CAB}, \tag{6.90}$$

or

$$\widehat{BCA} = 2\widehat{ABC}, \tag{6.91}$$

or

$$\widehat{ABC} = 2\widehat{CAB}. \tag{6.92}$$

Here, in order to simplify the computations, we use the law of cosines and we have that

$$\cos \widehat{CAB} = \sqrt{\frac{1+\cos \widehat{BCA}}{2}} \tag{6.93}$$

in the case of (6.90).

In the case of (6.91), we get

$$\cos \widehat{ABC} = \sqrt{\frac{1+\cos \widehat{BCA}}{2}}, \tag{6.94}$$

and finally in the case of (6.92), we obtain

$$\cos \widehat{CAB} = \sqrt{\frac{1+\cos \widehat{ABC}}{2}}. \tag{6.95}$$

As we have already seen by virtue of the relations (6.83), (6.84), and (6.85), the above cosines are rational numbers. Therefore, the quantities

$$(1+\cos \widehat{ABC})/2 \quad \text{and} \quad (1+\cos \widehat{BCA})/2$$

are perfect squares of fractions.

If $b = 2$, then

$$\cos \widehat{BCA} = -1, \tag{6.96}$$

which is impossible. In the respective cases for $b = 3, 4, 5, 6$, we have

$$\frac{1+\cos \widehat{ABC}}{2} = \frac{27}{32}, \frac{4}{5}, \frac{25}{32}, \frac{27}{35}, \tag{6.97}$$

respectively, and thus

$$\frac{1+\cos \widehat{BCA}}{2} = \frac{3}{8}, \frac{1}{2}, \frac{9}{16}, \frac{3}{5}, \tag{6.98}$$

respectively. Only in the case when $b = 5$ we have

$$\cos \widehat{CAB} = \frac{3}{4}, \tag{6.99}$$

which means that

$$\widehat{BCA} = 2\widehat{CAB} \tag{6.100}$$

and therefore $b = 5$, $c = 6$, and $a = 4$.

We have reached the conclusion that there is a unique triangle with side lengths consecutive integers and such that one of its angles is twice as large as another of its angles. Therefore, the triangles ABC and $A_1B_1C_1$ are equal, and therefore have equal areas. □

Proposition 6.1 *In a triangle ABC inscribed in a circle (O, R), the center I of the inscribed circle in the triangle ABC is determined by the intersection of the circle (D, DB), where D is the midpoint of the arc defined by the points B, C such that the vertex A does not belong to this arc.*

Proof Let I be the intersection point of the bisectors of the triangle ABC (i.e., I is the center of the inscribed circle), the bisector of the angle $\widehat{A}$ passes through the midpoint of the arc BC, with A not belonging to this arc. Then

$$\widehat{IBD} = \frac{\widehat{B}}{2} + \widehat{DBC} = \frac{\widehat{B}}{2} + \frac{\widehat{A}}{2} = \widehat{BID}, \tag{6.101}$$

which implies that $DI = DB$. Similarly, we get $DI = DC$.

For the converse, let us denote by I the common point of the bisector of the angle $\widehat{A}$ with the circle (D, DB), where $DB = DC$. Let D be the midpoint of the arc BC such that A is not in this arc. Then

$$\widehat{IBC} = \frac{\widehat{IDC}}{2} = \frac{\widehat{B}}{2}, \tag{6.102}$$

since the half-line BI coincides with the bisector of the angle $\widehat{B}$.

Similarly, we can verify that the half-line CI is the bisector of the angle $\widehat{C}$. Hence, the point I is actually the point of intersection of the bisectors of the triangle ABC, that is, *the center of the circle inscribed in the triangle ABC.* □

6.2.3 Let a triangle ABC be given. Investigate the possibility of determining a point M in the interior of ABC such that if D, E, Z are the projections of M to the sides AB, BC, CA, respectively, then the relations

$$\frac{AD}{m} = \frac{BE}{n} = \frac{CZ}{l} \tag{6.103}$$

should hold, if m, n and l are lengths of given line segments.

Solution With no loss of generality, we may assume that $m < a$, $n < c$, and $l < b$, where a, b, c are the lengths of the sides BC, CA, and AB of the triangle ABC. We proceed by using the method of *Proof by Analysis* (see Fig. 6.23). Let M be a point in the interior of the triangle ABC having the desired properties. We observe that

$$\frac{AD}{m} = \frac{BE}{n} \quad \Leftrightarrow \quad mBE + nDB = cn. \tag{6.104}$$

On the half-line BA, we take the line segment $BH = n$, and on the half-line BC, the point Q such that $BQ = m$. We consider the parallelogram $HBQB'$. We are going to use the following lemma:

Lemma 6.4 *Let $ABCD$ be a parallelogram and let a circle passing through its vertex A meet the sides AB, AD at the points Z, E and the diagonal AC at the point H.*

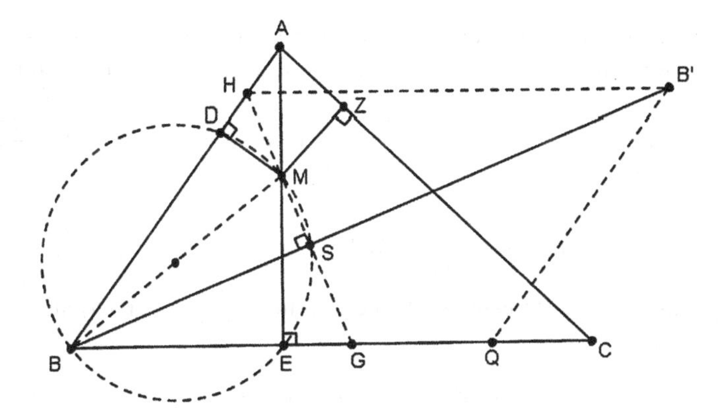

Fig. 6.23 Illustration of Problem 6.2.3

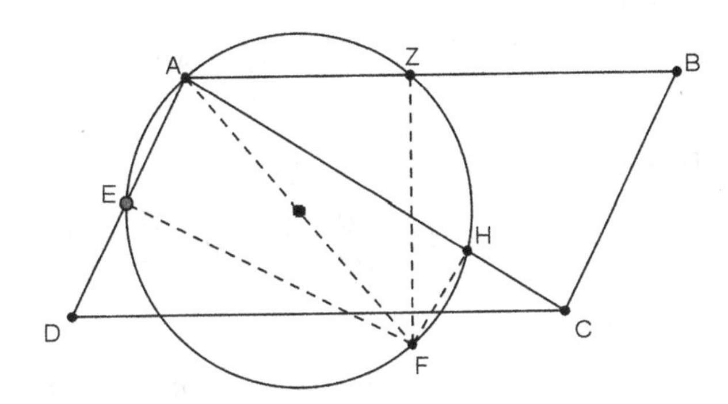

Fig. 6.24 Illustration of Problem 6.2.3 (Lemma 6.4)

Then the following relation holds (*see Fig.* 6.24):

$$AB \cdot AZ + AD \cdot AE = AH \cdot AC. \tag{6.105}$$

Let S be the projection of the point M to the diagonal BB'. It is a fact that the points B, E, S, M, D are homocyclic and belong to a circle of diameter BM since

$$\widehat{MEB} = \widehat{MSB} = \widehat{BDM} = 90^\circ.$$

Based on Lemma 6.4 and (6.105), we deduce

$$BD \cdot BH + BE \cdot BQ = BS \cdot BB' = c \cdot n \quad \Rightarrow \quad BS = \frac{c \cdot n}{BB'}. \tag{6.106}$$

The relation (6.106) implies that S is a fixed point and that the point M belongs to the perpendicular to the constant line BB' at its point S. These facts are valid with respect to the diagonal BB'. If we apply the same method for the corresponding diagonal CC', we obtain that the point M belongs to another line. The intersection of these lines determines the point M. We have thus proved that such a point M actually exists.

Remark We can consider $m < a$, $n < c$, and $l < b$ since by (6.104) a real number $t \in \mathbb{R}$ can be determined in such a way that the line segments satisfy the relations

$$m_1 = tm, \qquad n_1 = tn \quad \text{and} \quad l_1 = tl, \quad \text{with } tm < a, \text{ that is, } t < \frac{a}{m},$$

$$t \cdot n < c, \quad \text{that is, } t < \frac{c}{n}, \quad \text{and} \quad t \cdot l < b, \quad \text{that is, } t < \frac{b}{l}.$$

By an application of the Ptolemy's Theorem to the inscribed quadrilateral $AEHZ$, we get

$$AE \cdot HZ + AZ \cdot EH = AH \cdot EZ.$$

Note that $\widehat{ZHE} = 180^\circ - \widehat{A}$ and $\widehat{EZH} = \widehat{EAH}$. Therefore, the triangles HZE and ADC are similar. Consequently,

$$\frac{CB}{ZH} = \frac{AD}{HZ} = \frac{AB}{EH} = \frac{AC}{EZ}.$$

Thus

$$AE \cdot HZ \cdot \frac{AD}{HZ} + AZ \cdot EH \cdot \frac{AB}{EH} = AH \cdot EZ \cdot \frac{AC}{EZ}.$$

Therefore,

$$AE \cdot AD + AZ \cdot AB = AH \cdot AC,$$

that is,

$$AB \cdot AZ + AD \cdot AE = AH \cdot AC.$$

□

Remark The relation

$$AB \cdot AZ + AD \cdot AE = AH \cdot AC$$

can be proved by using the inner product of vectors. It is enough to consider the point F antidiametrical to the point A and to observe that

$$AB \cdot AZ + AD \cdot AE = \overrightarrow{AB} \cdot \overrightarrow{AZ} + \overrightarrow{AD} \cdot \overrightarrow{AE}$$

with

$$\begin{aligned} \overrightarrow{AB} \cdot \overrightarrow{AZ} + \overrightarrow{AD} \cdot \overrightarrow{AE} &= \overrightarrow{AB} \cdot \overrightarrow{AF} + \overrightarrow{AD} \cdot \overrightarrow{AF} \\ &= (\overrightarrow{AB} + \overrightarrow{AD}) \cdot \overrightarrow{AF} \\ &= \overrightarrow{AC} \cdot \overrightarrow{AF}. \end{aligned}$$

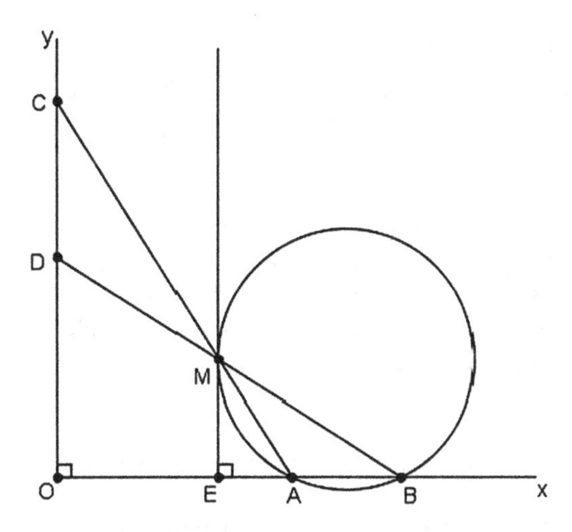

Fig. 6.25 Illustration of Problem 6.2.4

Hence

$$\begin{aligned} AB \cdot AZ + AD \cdot AE &= \overrightarrow{AC} \cdot \overrightarrow{AF} \\ &= \overrightarrow{AC} \cdot \overrightarrow{AH}, \end{aligned}$$

and thus

$$AB \cdot AZ + AD \cdot AE = AC \cdot AH. \qquad \square$$

6.2.4 Let $\widehat{xOy}$ be a right angle and on the side Ox fix two points A, B with $OA < OB$. On the side Oy, we consider two moving points C, D such that $OD < OC$ with $CD/DO = m/n$, where m, n are given positive integers. If M is the point of intersection of AC and BD, determine the position of M under the assumption that the angle $\widehat{DMA}$ attains its minimum.

Solution It should be enough to determine the maximum attained by the angle $\widehat{AMB}$. Using the theorem of Menelaus, we have (see Fig. 6.25)

$$\frac{CD}{CO} \cdot \frac{OA}{AB} \cdot \frac{BM}{MD} = 1,$$

which implies

$$\frac{BM}{MD} = \frac{(m+n)AB}{m \cdot OA} = \frac{BE}{EO}, \tag{6.107}$$

where $E \in Ox$ with $ME \perp Ox$.

However, by the theorem of Thales, we also have

$$\frac{BM}{MD} = \frac{BE}{EO}, \tag{6.108}$$

and thus

$$\frac{BE}{EO} = \frac{(m+n)AB}{m \cdot OA}. \tag{6.109}$$

This relation implies that the perpendicular straight line ME to the side Ox at the point E is constant. It follows that the point $M \in \epsilon$ such that $\widehat{AMB}$ attains a maximum occurs when the prescribed circle of the triangle MAB obtains its minimal radius, that is, when passing through the points A and B, and is tangential to the line ϵ at a point M. This point M is completely determined (and thus constructed) from the relation

$$EM^2 = EA \cdot EB.$$

This is known as the *Apollonius' construction.* □

Note 1 The problem of the determination of the point M on the line ϵ such that the line segment AB forms an angle $\widehat{AMB}$ that becomes maximum, when the points A, B belong to the same half-plane determined by the straight line ϵ, is traditionally called the *statue problem.*

6.2.5 Given $\widehat{xOy} = 60^\circ$, we consider the points A, B moving on the sides Ox and Oy, respectively, so that the length of the line segment AB is preserved subject to the assumption that the triangle OAB is not an obtuse triangle. Let D, E, Z be the feet of the heights OD, AE, and BZ of the triangle OAB to AB, BO, and OA, respectively. Compute the maximal value of the sum

$$\sqrt{DE} + \sqrt{EZ} + \sqrt{ZD}.$$

Solution By assumption, the length of the straight line segment AB is constant and also the angle $\widehat{xOy} = 60^\circ$ is constant, therefore the circle determined by the triangle OAB is of constant radius R since the isosceles triangle KAB has its basis AB of constant length and its angle $\widehat{AKB} = 120^\circ$, when the point K is the center of the circle (OAB) (see Fig. 6.26). We observe that for the areas of the triangles, one has

$$S_{OAB} = S_{OZKE} + S_{ADKZ} + S_{DBEK}.$$

Hence, recalling that $OK = AK = BK = R$, we obtain

$$S_{OAB} = \frac{ZE \cdot R}{2} + \frac{ZD \cdot R}{2} + \frac{DE \cdot R}{2}. \tag{6.110}$$

By (6.110) and the Cauchy–Schwarz–Buniakowski inequality, we deduce

$$(\sqrt{DE} + \sqrt{EZ} + \sqrt{ZD})^2 \leq \left[\left(\frac{1}{\sqrt{R}}\right)^2 + \left(\frac{1}{\sqrt{R}}\right)^2 + \left(\frac{1}{\sqrt{R}}\right)^2\right] \cdot \left[(\sqrt{ZE \cdot R})^2 + (\sqrt{ZD \cdot R})^2 + (\sqrt{DE \cdot R})^2\right]$$

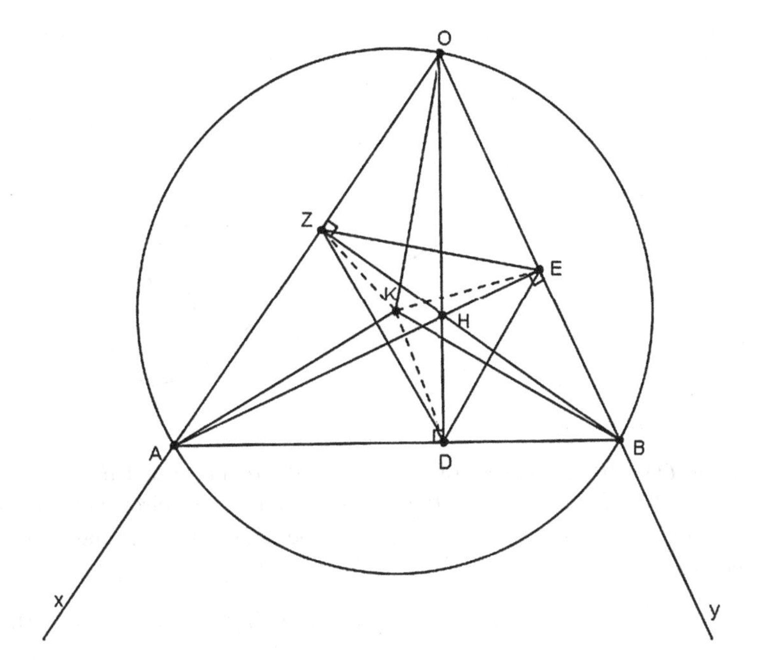

Fig. 6.26 Illustration of Problem 6.2.5

$$= \frac{3}{2} \cdot 2S_{OAB} \leq \frac{3\sqrt{3}R^2}{4}. \tag{6.111}$$

Thus, the maximal value of the sum

$$\sqrt{DE} + \sqrt{EZ} + \sqrt{ZD}$$

is equal to

$$3\sqrt{\frac{R\sqrt{3}}{2}},$$

which is achieved for $OA = OB$.

Remark 6.3 Study the same problem in the case $\widehat{xOy} \neq 60°$. □
(*Open problem.*)

6.2.6 Let O be a given point outside a given circle of center C. Let OPQ be any secant of the circle passing through O and R be a point on PQ such that

$$\frac{OP}{QO} = \frac{PR}{RQ}.$$

Find the geometrical locus of the point R.

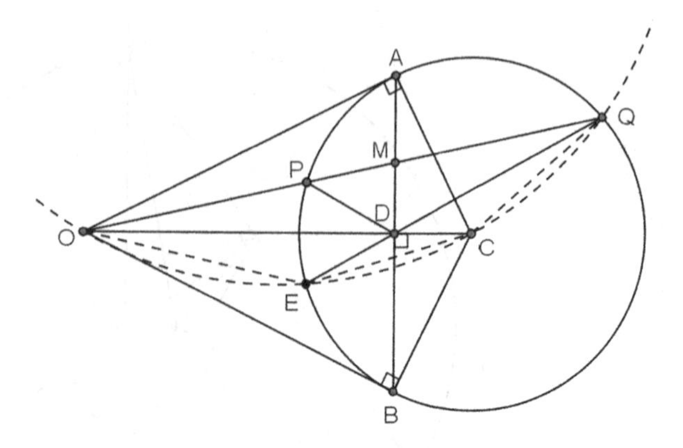

Fig. 6.27 Illustration of Problem 6.2.6

Solution Let OA, OB be tangents of the circle at the points A and B, respectively. Let $D = AB \cap OC$ and $M = PQ \cap AB$. Extend QD to meet the circle at the point E. Since the points O, B, C, A are concyclic and the points A, E, B and Q are also concyclic, we obtain

$$OD \cdot DC = AD \cdot DB = ED \cdot DQ, \tag{6.112}$$

and thus the points O, Q, C, E are concyclic (see Fig. 6.27). Therefore, we have

$$\widehat{COQ} = \widehat{QEC} = \widehat{EQC} = \widehat{COE}, \tag{6.113}$$

and so P, E are mirror images with respect to the straight line OC. Therefore,

$$\widehat{PDO} = \widehat{EDO}, \tag{6.114}$$

that is, OD is the bisector of the external angle of $\widehat{PDQ}$ of the triangle PDQ. Hence we have

$$\frac{PD}{DQ} = \frac{PM}{MQ}. \tag{6.115}$$

Combining the above relations, we deduce

$$\frac{OP}{OQ} = \frac{PM}{MQ}, \tag{6.116}$$

and therefore $M = R$. Hence, the geometrical locus of R is the straight line segment AB. □

6.2.7 Prove that in each triangle the following equality holds:

$$\frac{1}{r}\left(\frac{b^2}{r_b} + \frac{c^2}{r_c}\right) - \frac{a^2}{r_b r_c} = 4\left(\frac{R}{r_a} + 1\right), \tag{6.117}$$

where s is the semiperimeter of the triangle, S is the area enclosed by the triangle, a, b, c are the sides of the triangle, R is the radius of the circumscribed circle, r is the corresponding radius of the inscribed circle, and r_a, r_b, r_c are the radii of the corresponding exscribed circles of the triangle.

(*Proposed by Dorin Andrica, Romania and Khoa Lu Nguyen [14], USA*)

Solution (by Prithwijit De, Calcutta, India) We have

$$r_a = \frac{S}{s-a}, \qquad r_b = \frac{S}{s-b}, \qquad r_c = \frac{S}{s-c}, \qquad r = \frac{S}{s}, \qquad R = \frac{abc}{4S}. \tag{6.118}$$

Thus, we obtain (see Fig. 6.28)

$$\begin{aligned}
&\frac{1}{r}\left(\frac{b^2}{r_b} + \frac{c^2}{r_c}\right) - \frac{a^2}{r_b r_c} \\
&= \frac{b^2 s(s-b) + c^2 s(s-c) - a^2(s-b)(s-c)}{S^2} \\
&= \frac{(b^2+c^2-a^2)s^2 - s(b+c)(b^2-bc+c^2) + a^2(b+c)s - a^2bc}{S^2} \\
&= \frac{s(b^2+c^2-a^2)(s-b-c) + bc((b+c)s - a^2)}{S^2} \\
&= \frac{2a^2(b^2+c^2+bc) - a^4 - (b^4+c^4-2b^2c^2) + 2abc(b+c) - 4a^2bc}{4S^2} \\
&= \frac{2(a^2b^2+b^2c^2+c^2a^2) - a^4 - b^4 - c^4 + 2abc(b+c-a)}{4S^2} \\
&= \frac{16S^2 + 4abc(s-a)}{4S^2} = 4 + \frac{4SR(s-a)}{S^2} \\
&= 4\left(\frac{R}{r_a} + 1\right).
\end{aligned} \tag{6.119}$$

This completes the proof. □

6.2.8 Let $A_1A_2A_3A_4A_5$ be a convex planar pentagon and let $X \in A_1A_2$, $Y \in A_2A_3$, $Z \in A_3A_4$, $U \in A_4A_5$, and $V \in A_5A_1$ be points such that A_1Z, A_2U, A_3V, A_4X, A_5Y intersect at the point P. Prove that

$$\frac{A_1X}{A_2X} \cdot \frac{A_2Y}{A_3Y} \cdot \frac{A_3Z}{A_4Z} \cdot \frac{A_4U}{A_5U} \cdot \frac{A_5V}{A_1V} = 1. \tag{6.120}$$

(*Proposed by Ivan Borsenko [26], USA*)

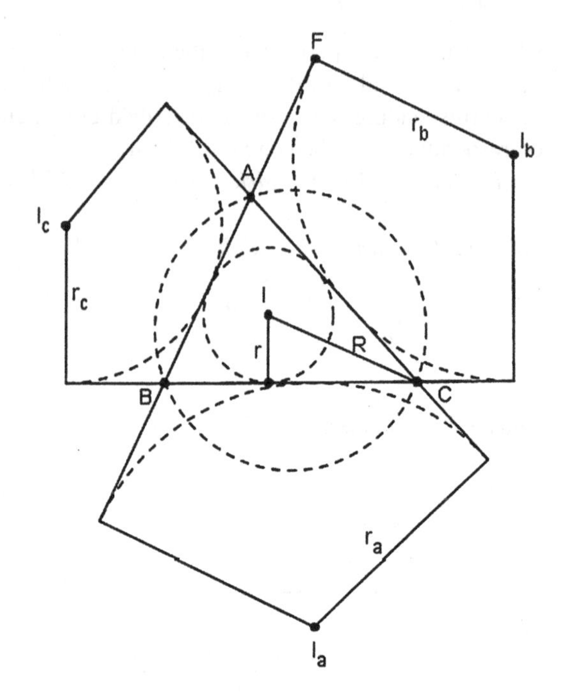

Fig. 6.28 Illustration of Problem 6.2.7

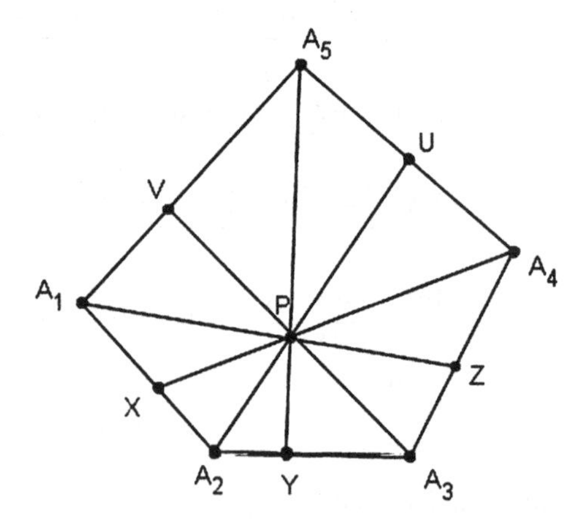

Fig. 6.29 Illustration of Problem 6.2.8

Solution (by Ercole Suppa, Italy) We shall make use of the following (see Fig. 6.29)

Lemma 6.5 *If P is a point on the side BC of a triangle ABC then*

$$\frac{PB}{PC} = \frac{AB}{AC} \cdot \frac{\sin \widehat{PAB}}{\sin \widehat{PAC}}. \tag{6.121}$$

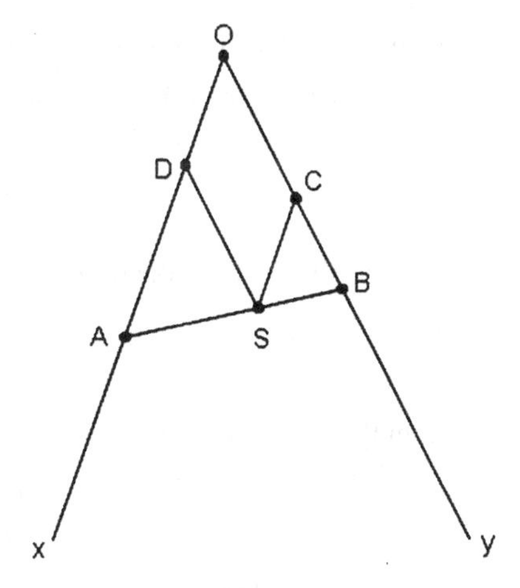

Fig. 6.30 Illustration of Problem 6.2.9

Let us denote

$$\widehat{A_1PX} = \widehat{A_4PZ} = \alpha,$$

$$\widehat{XPA_2} = \widehat{UPA_4} = \beta,$$

$$\widehat{A_2PY} = \widehat{A_5PU} = \gamma,$$

$$\widehat{YPA_3} = \widehat{VPA_5} = \delta,$$

and

$$\widehat{A_3PZ} = \widehat{A_1PV} = \epsilon.$$

From the above Lemma 6.5 applied to the triangles A_1PA_2, A_2PA_3, A_3PA_4, A_4PA_5, and A_5PA_1, we obtain

$$\begin{aligned}\frac{A_1X}{A_2X} \cdot \frac{A_2Y}{A_3Y} \cdot \frac{A_3Z}{A_4Z} \cdot \frac{A_4U}{A_5U} \cdot \frac{A_5V}{A_1V} &= \frac{\sin\alpha}{\sin\beta} \cdot \frac{\sin\gamma}{\sin\delta} \cdot \frac{\sin\alpha}{\sin\beta} \cdot \frac{\sin\epsilon}{\sin\alpha} \cdot \frac{\sin\beta}{\sin\gamma} \cdot \frac{\sin\delta}{\sin\epsilon}\\ &= 1, \end{aligned} \tag{6.122}$$

and thus (6.120) is proved. □

6.2.9 Given an angle $\widehat{xOy}$ and a point S in its interior, consider a straight line passing through S and intersecting the sides Ox, Oy at the points A and B, respectively. Determine the position of AB so that the product $OA \cdot OB$ attains its minimum.

Solution Let D, C be points on the sides Ox, Oy, respectively, such that the straight line SD is parallel to the straight line Oy and the straight line SC is parallel to the straight line Ox. It follows that the straight line segments $OD = b$ and $OC = a$ are constant (see Fig. 6.30).

The problem can be formulated equivalently as follows:

Determine the position of AB so that the product

$$\frac{a}{OA} \cdot \frac{b}{OB}$$

becomes maximum.

From the similarity of the triangles OAB, CSB, it follows that

$$\frac{a}{OA} = \frac{BS}{BA}, \tag{6.123}$$

and from the similarity of the triangles OAB and DAS, we get

$$\frac{b}{OB} = \frac{AS}{BA}. \tag{6.124}$$

Adding (6.123) and (6.124), we obtain

$$\frac{a}{OA} + \frac{b}{OB} = 1. \tag{6.125}$$

It is a fact that when the sum of two positive real numbers is constant, their product attains its maximal value when the numbers are equal. Therefore,

$$\frac{a}{OA} = \frac{b}{OB}.$$

This means that *the straight lines AB and DC are parallel.* □

6.2.10 Let the incircle of a triangle ABC touch the sides BC, CA, AB at the points D, E, F, respectively. Let K be a point on the side BC and M be the point on the line segment AK such that $AM = AE = AF$. Denote by L and N the incenters of the triangles ABK and ACK, respectively. Prove that K is the foot of the altitude from A if and only if $DLMN$ is a square.

(*Proposed by Bogdan Enescu [41], Romania*)

Solution (by Ercole Suppa, Teramo, Italy) We will first prove the following two lemmas (see Fig. 6.31):

Lemma 6.6 *The points D, K lie on the circle with diameter LN.*

Proof Without loss of generality, let $c < b$. Assume I is the incenter of the triangle ABC and U, V are the points where the circles (L), (N) touch the side BC. Let r, r_1, r_2 be the inradii of the circles (I), (L), (N) as shown in the figure. Let $a = BC$, $b = CA$, $c = AB$, $m = BK$, $n = KC$, $x = AK$.

Because of the fact that L, N are the incenters of the triangles ABK and ACK, we get

$$\widehat{LKN} = \widehat{LKA} + \widehat{AKN} = \frac{1}{2}(\widehat{BKA} + \widehat{AKC}) = 90^\circ. \tag{6.126}$$

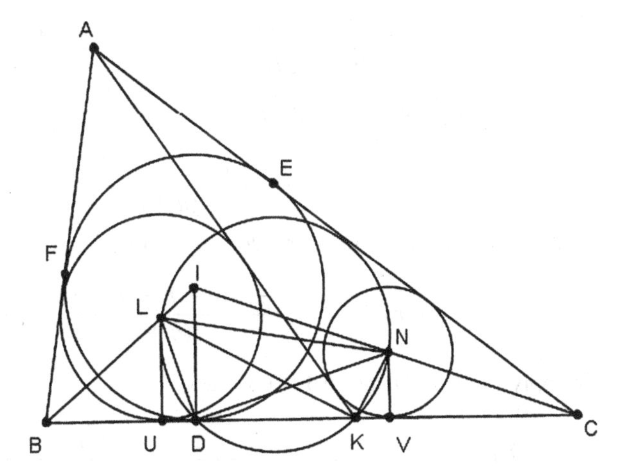

Fig. 6.31 Illustration of Problem 6.2.10

To prove that $\widehat{LDN} = 90°$, it is sufficient to show that

$$LD^2 + DN^2 = LN^2. \tag{6.127}$$

From the theorem of Pythagoras, we obtain

$$LD^2 = r_1^2 + UD^2, \tag{6.128}$$

$$ND^2 = r_2^2 + DV^2, \tag{6.129}$$

$$\begin{aligned} LN^2 &= UV^2 + (r_1 - r_2)^2 \\ &= UD^2 + DV^2 + 2UD \cdot DV + r_1^2 + r_2^2 - 2r_1r_2. \end{aligned} \tag{6.130}$$

To prove (6.127), it is sufficient to show that

$$UD \cdot DV = r_1r_2.$$

We have

$$UD = BD - BU = \frac{a+c-b}{2} - \frac{m+c-x}{2} = \frac{a+x-b-m}{2}, \tag{6.131}$$

$$DV = DC - CV = \frac{a+b-c}{2} - \frac{n+b-x}{2} = \frac{a+x-c-n}{2}. \tag{6.132}$$

By (6.131) and (6.132), and by setting $n = a - m$, we get

$$UD \cdot DV = \frac{(x+a-b-m)(x-c+m)}{4}. \tag{6.133}$$

From the similarity of the triangles BUL, BDI, and CVN, CDI, we deduce

$$\frac{LU}{ID} = \frac{BU}{BD} \quad \Rightarrow \quad r_1 = r \cdot \frac{c+m-x}{a+c-b}, \tag{6.134}$$

$$\frac{NV}{ID} = \frac{CV}{CD} \quad \Rightarrow \quad r_2 = r \cdot \frac{b+n-x}{a+b-c}. \tag{6.135}$$

From the above equalities and since

$$r^2 = \frac{(b+c-a)(a+c-b)(a+b-c)}{4(a+b+c)}$$

(it is left as an exercise to the reader), by setting $n = a - m$, we derive

$$r_1 r_2 = \frac{(b+c-a)(a+b-m-x)(c+m-x)}{4(a+b+c)}. \tag{6.136}$$

By applying (6.133) and (6.136), we get

$$UD \cdot DV - r_1 r_2 = \frac{ax^2 - ac^2 + a^2m - b^2m + c^2m - am^2}{2(a+b+c)}. \tag{6.137}$$

From Stewart's theorem, we obtain

$$x^2 = \frac{mb^2 + (a-m)c^2 - am(a-m)}{a}. \tag{6.138}$$

By substituting x^2 from (6.138) into (6.137) and carrying out the calculations (it is left as an exercise to the reader), we obtain

$$UD \cdot DV - r_1 r_2 = \frac{(c^2 - am)(m+n-a)}{2(a+b+c)} = 0. \tag{6.139}$$

Therefore,

$$LD^2 + DN^2 = LN^2,$$

and this completes the proof of the lemma. □

Lemma 6.7 *Let M be the second intersection point of AK with the circle γ circumscribed to the quadrilateral $DKNL$. Then $DM \perp LN$ and*

$$AM = AE = AF.$$

Proof Assume that the incircle of the triangle ABK intersects the side AB at the point F'. According to Lemma 6.6, the center of γ is the midpoint of LN, and therefore the point M lies on the external tangent to the circles (L), (N). Therefore, it follows that $DM \perp LN$, and thus

$$\begin{aligned} AM &= AF' - UD = AF' - (BD - BU) \\ &= \frac{c+x-m}{2} - \frac{a+c-b}{2} + \frac{c+m-x}{2} \\ &= \frac{b+c-a}{2} = AF. \end{aligned} \tag{6.140}$$

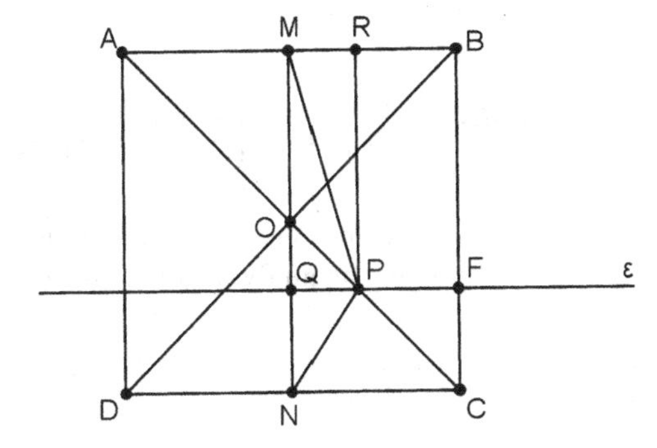

Fig. 6.32 Illustration of Problem 6.2.11

This completes the proof of Lemma 6.7. □

Using Lemmas 6.6 and 6.7, it follows that

- $DLMN$ is cyclic;
- $\widehat{LDN} = \widehat{LMN} = 90°$;
- $DM \perp LN$.

Hence the quadrilateral $DLMN$ is a square if and only if MD is a diameter of the circumcircle of $DLMN$, that is, $\widehat{MKD} = 90°$. Thus $AK \perp BC$. □

6.2.11 Let $ABCD$ be a square of center O. The parallel through O to AD intersects AB and CD at the points M and N, respectively, and a parallel to AB intersects the diagonal AC at the point P. Prove that

$$OP^4 + \left(\frac{MN}{2}\right)^4 = MP^2 \cdot NP^2. \tag{6.141}$$

(Proposed by Titu Andreescu [7], USA)

Solution (by Christopher Wiriawan, Indonesia) Let Q be the intersection of the straight line MN and the parallel ϵ to AB passing through P. Let R be the foot of the perpendicular from P to AB. Then $QN = RB$ because of the fact that (see Fig. 6.32)

$$QN = FC = FP = BR,$$

and thus, by Pythagoras' theorem, we deduce

$$\begin{aligned} OP^4 + \left(\frac{MN}{2}\right)^4 &= (OQ^2 + QP^2)^2 + \left(\frac{MN}{2}\right)^4 \\ &= \left(\left(\frac{MN}{2} - QN\right)^2 + \left(\frac{MN}{2} - RB\right)^2\right)^2 + \left(\frac{MN}{2}\right)^4. \end{aligned} \tag{6.142}$$

This is equivalent to

$$\left(2\left(\frac{MN}{2}-QN\right)^2\right)^2+\left(\frac{MN}{2}\right)^4=4\left(\frac{MN}{2}-QN\right)^4+\left(\frac{MN}{2}\right)^4. \tag{6.143}$$

It suffices to prove that the above expression is equal to the right-hand side of (6.141). By applying again Pythagoras' theorem, we have

$$\begin{aligned} NP^2 &= QN^2+QP^2=QN^2+\left(\frac{MN}{2}-QN\right)^2 \\ &= 2QN^2-MN\cdot QN+\left(\frac{MN}{2}\right)^2. \end{aligned} \tag{6.144}$$

Also

$$\begin{aligned} MP^2 &= QP^2+MQ^2=\left(\frac{MN}{2}-QN\right)^2+(MN-QN)^2 \\ &= 5\left(\frac{MN}{2}\right)^2-3MN\cdot QN+2QN^2. \end{aligned} \tag{6.145}$$

Hence

$$\begin{aligned} NP^2\cdot MP^2 = &\left(2QN^2-MN\cdot QN+\left(\frac{MN}{2}\right)^2\right) \\ &\cdot\left(5\left(\frac{MN}{2}\right)^2-3MN\cdot QN+2QN^2\right). \end{aligned} \tag{6.146}$$

This is equivalent to

$$\begin{aligned} 4QN^4 = {} & 16QN^3\left(\frac{MN}{2}\right)+24QN^2\left(\frac{MN}{2}\right)^2 \\ & -16QN\left(\frac{MN}{2}\right)^3+4\left(\frac{MN}{2}\right)^4+\left(\frac{MN}{2}\right)^4, \end{aligned} \tag{6.147}$$

that is,

$$4\left(\frac{MN}{2}-QN\right)^4+\left(\frac{MN}{2}\right)^4.$$

This completes the proof. □

Second solution Denote $OM=a$ and $OQ=x$, then we deduce that

$$OP=x\sqrt{2},\qquad MP^2=(a+x)^2+x^2,\qquad NP^2=(a-x)^2+x^2.$$

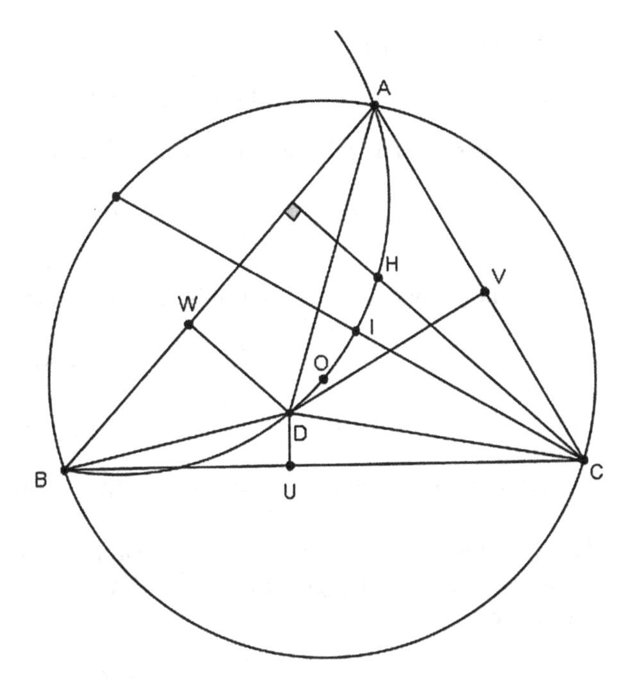

Fig. 6.33 Illustration of Problem 6.2.12

Therefore,

$$\begin{aligned} MP^2 \cdot NP^2 &= (2x^2 + a^2 + 2ax)(2x^2 + a^2 - 2ax) \\ &= (2x^2 + a^2)^2 - 4a^2x^2 \\ &= 4x^2 + a^4 \\ &= OP^4 + \left(\frac{MN}{2}\right)^4. \end{aligned}$$

□

6.2.12 Let O, I, H be the circumcenter, the incenter and the orthocenter of the triangle ABC, respectively, and let D be a point in the interior of ABC such that

$$BC \cdot DA = CA \cdot DB = AB \cdot DC.$$

Prove that the points A, B, D, O, I, H are concyclic if and only if $\widehat{C} = 60°$.
(Proposed by T. Andreescu (USA), D. Andrica and C. Barbu [8] (Romania))

Solution (by Daniel Lasaosa, Spain) Let U, V, W be the projections of the point D onto BC, CA, AB. Then, it can be proved that UVW is an equilateral triangle and indeed it holds (see Fig. 6.33)

$$\widehat{ADB} = \widehat{C} = 60°.$$

Because of the fact that

$$\widehat{AWD} = \widehat{AVD} = 90^\circ,$$

the quadrilateral $AVDW$ is cyclic with diameter DA, or

$$VW = AD \sin \widehat{A} = \frac{BC \cdot AD}{2R}$$

and by cyclic permutation of A, B, C this quantity is equal to $UV \cdot WD$. Furthermore,

$$\widehat{ADW} = \widehat{AVW} = 180^\circ - \widehat{A} - \widehat{AWV},$$

as well as

$$\widehat{ADB} = \widehat{ADW} + \widehat{BDW} = 360^\circ - \widehat{A} - \widehat{B} - \widehat{AWV} - \widehat{BWU} = \widehat{C} + 60^\circ. \qquad (6.148)$$

The point D is called the *first isodynamic point*. It is inside the triangle ABC if and only if no angle of the triangle ABC exceeds 120°.

We have

$$\widehat{AIB} = 90^\circ + \frac{1}{2}\widehat{C}$$

and

$$\widehat{AHB} = 180^\circ - \widehat{C}.$$

If ABC is an obtuse triangle at C, then O, C are on opposite sides of the side AB and

$$\widehat{AOB} = 360^\circ - 2\widehat{C}.$$

However, if ABC is not an obtuse triangle at C, then O, C are on the same side of the side AB and

$$\widehat{AOB} = 2\widehat{C}.$$

Therefore, A, B, O, I are cocyclic if and only if the triangle ABC is acute at C, otherwise we would need

$$\widehat{AOB} + \widehat{AIB} = 180^\circ,$$

or equivalently,

$$270^\circ = \frac{3}{2}\widehat{C},$$

that is, $\widehat{C} = 180^\circ$, which is absurd since ABC would be degenerate and O, I could not be defined.

Thus we can assume that the triangle $\widehat{ABC}$ has an acute angle at the vertex C.

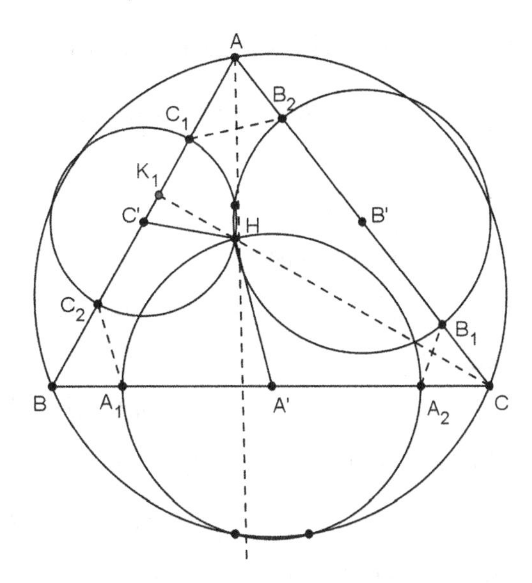

Fig. 6.34 Illustration of Problem 6.2.13

If the triangle ABC is acute at C, then A, B and any two of the points D, O, I, H are concyclic if and only if the corresponding pair from the following four angles are equal, that is,

$$\widehat{AIB} = 90 + \frac{\widehat{C}}{2}, \qquad \widehat{AHB} = 180^\circ - \widehat{C},$$
$$\widehat{AOB} = 2\widehat{C}, \qquad \widehat{ADB} = \widehat{C} + 60^\circ, \tag{6.149}$$

that is, if and only if $\widehat{C} = 60^\circ$. This completes the proof. □

6.2.13 Let H be the orthocenter of an acute triangle ABC and let A', B', C' be the midpoints of the sides BC, CA, AB, respectively. Denote by A_1 and A_2 the intersections of the circle $(A', A'H)$ with the side BC. In the same way, we define the points B_1, B_2 and C_1, C_2, respectively. Prove that the points A_1, A_2, B_1, B_2, C_1, C_2 are cocyclic.

(*Proposed by Catalin Barbu [24], Romania*)

Solution (by Michel Bataille, France) The power of A_1 with respect to the prescribed circumcircle (O, R) of the triangle ABC is (see Fig. 6.34)

$$OA_1^2 - R^2 = \left(A_1A_1'\right)^2 - \frac{BC^2}{4}. \tag{6.150}$$

If K_1 is the orthogonal projection of C onto the side AB, we get

$$\widehat{BCK_1} = 90^\circ - \widehat{B}.$$

From the law of cosines applied to the triangle CHA', we obtain

$$\begin{aligned} A'H^2 &= A'C^2 + CH^2 - 2A'C \cdot CH \cos\left(90^\circ - \widehat{B}\right) \\ &= \frac{BC^2}{4} + 4\left(OC'\right)^2 - 2BC \cdot OC' \sin \hat{B}. \end{aligned} \tag{6.151}$$

We have used the property

$$CH = 2OC'.$$

Since

$$OC' = R \cos \widehat{C}$$

($\widehat{C}$ is acute) and

$$BC = 2R \sin A, \qquad A'H = A_1A',$$

the relations (6.150) and (6.151) yield

$$OA_1^2 = R^2 + 4R^2 \cos \widehat{C}(\cos \widehat{C} - \sin \widehat{A} \sin \widehat{B}). \tag{6.152}$$

But,

$$\cos \widehat{C} - \sin \widehat{A} \sin \widehat{B} = -\cos(A+B) - \sin A \sin B = -\cos A \cos B.$$

Thus

$$OA_1^2 = R^2(1 - 4\cos A \cos B \cos C). \tag{6.153}$$

Due to the symmetry of the result, we note that

$$OA_1 = OA_2 = OB_1 = OB_2 = OC_1 = OC_2.$$

Hence A_1, A_2, B_1, B_2, C_1, C_2 are all concyclic (with center O). □

6.2.14 Let ABC be a triangle with midpoints M_a, M_b, M_c of the sides BC, AC, AB, respectively. Let also X, Y, Z be the points of tangency of the incircle of the triangle $M_aM_bM_c$ with M_bM_c, M_cM_a and M_aM_b.

(a) Prove that the straight lines AX, BY, CZ are concurrent at some point P.
(b) If A_1, B_1, C_1 are points of the sides BC, AC, AB, respectively, such that the straight lines AA_1, BB_1, CC_1 are concurrent at the point P, then the perimeter of the triangle $A_1B_1C_1$ is greater than or equal to the semi-perimeter of the triangle ABC.

(*Proposed by Roberto Bosch Cabrera [34], Cuba*)

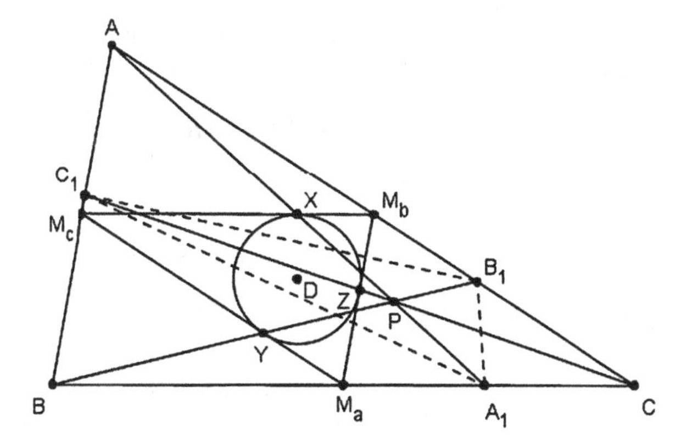

Fig. 6.35 Illustration of Problem 6.2.14

Solution (by Daniel Lasaosa, Spain) (a) By Thales' theorem and because of the fact that $M_cM_b \parallel BC$, we get (see Fig. 6.35)

$$\frac{BA_1}{A_1C} = \frac{M_cX}{XM_b}.$$

It can be easily proved that

$$M_cX = \frac{M_bM_c + M_cM_a - M_aM_b}{2} = \frac{a+b-c}{4} \tag{6.154}$$

and

$$XM_b = \frac{M_aM_b + M_bM_c - M_cM_a}{2} = \frac{a+c-b}{4}, \tag{6.155}$$

or

$$\frac{BA_1}{A_1C} = \frac{a+b-c}{c+a-b},$$

and similarly for its cyclic permutations. Applying the reciprocal of the Menelaus' theorem, we deduce that AX, BY, CZ meet at a point P. Since A_1 may be identified as the point where the side BC touches the excircle which touches the side BC and the extensions of the sides AB and AC, and similarly for B_1 and C_1, then the point P where AX, BY, CZ intersect is the Nagel's point of the triangle ABC.

(b) By the cosine law and Heron's formula for the area of the triangle ABC, we obtain

$$\begin{aligned} B_1C_1^2 &= AB_1^2 + AC_1^2 - 2AC_1 \cdot AB_1 \cos \widehat{A} \\ &= \frac{(a+b-c)^2}{4} + \frac{(a+c-b)^2}{4} \\ &\quad - \frac{(a+b-c)(a+c-b)(b^2+c^2-a^2)}{4bc} \\ &= a^2(1 - \sin B \sin C). \end{aligned} \tag{6.156}$$

Similarly, we get the formulas which correspond to its cyclic permutations.

However,

$$\begin{aligned} 2\sin B\sin C &= \cos(B-C)-\cos(B+C) \\ &\le 1+\cos A \\ &= 2-2\sin^2\frac{A}{2} \end{aligned} \tag{6.157}$$

and

$$B_1C_1 \ge a\sin\frac{A}{2}. \tag{6.158}$$

Hence, it suffices to prove that

$$\frac{a}{a+b+c}\sin\frac{A}{2}+\frac{b}{a+b+c}\sin\frac{B}{2}+\frac{c}{a+b+c}\sin\frac{C}{2}\ge\frac{1}{2}. \tag{6.159}$$

Because of the fact that

$$b+c=2R\cos\frac{A}{2}\cos\frac{B-C}{2}\le 2R\cos\frac{A}{2}, \tag{6.160}$$

by multiplying by $\sin^2\frac{A}{2}$, we obtain

$$(b+c)\sin^2\frac{A}{2}\le a\sin\frac{A}{2}. \tag{6.161}$$

Therefore,

$$\sum_{\text{cyclic}}(b+c)\sin^2\left(\frac{A}{2}\right)\le\sum_{\text{cyclic}}a\sin\frac{A}{2}, \tag{6.162}$$

but

$$\begin{aligned} \sum_{\text{cyclic}}(b+c)\sin^2\left(\frac{A}{2}\right) &= \frac{1}{2}\sum_{\text{cyclic}}(b+c)(1-\cos A) \\ &= 2s-\frac{1}{2}\sum_{\text{cyclic}}(b+c)\cos A \\ &= 2s-\frac{1}{2}\sum_{\text{cyclic}}(b+c+a)\cos A+\frac{1}{2}\sum_{\text{cyclic}}a\cos A \\ &= 2s-s\sum_{\text{cyclic}}\cos A+\frac{1}{2}\sum_{\text{cyclic}}a\cos A \\ &= 2s-s\left(1+\frac{r}{R}\right)+\frac{1}{2}\frac{2sr}{R}=s. \end{aligned} \tag{6.163}$$

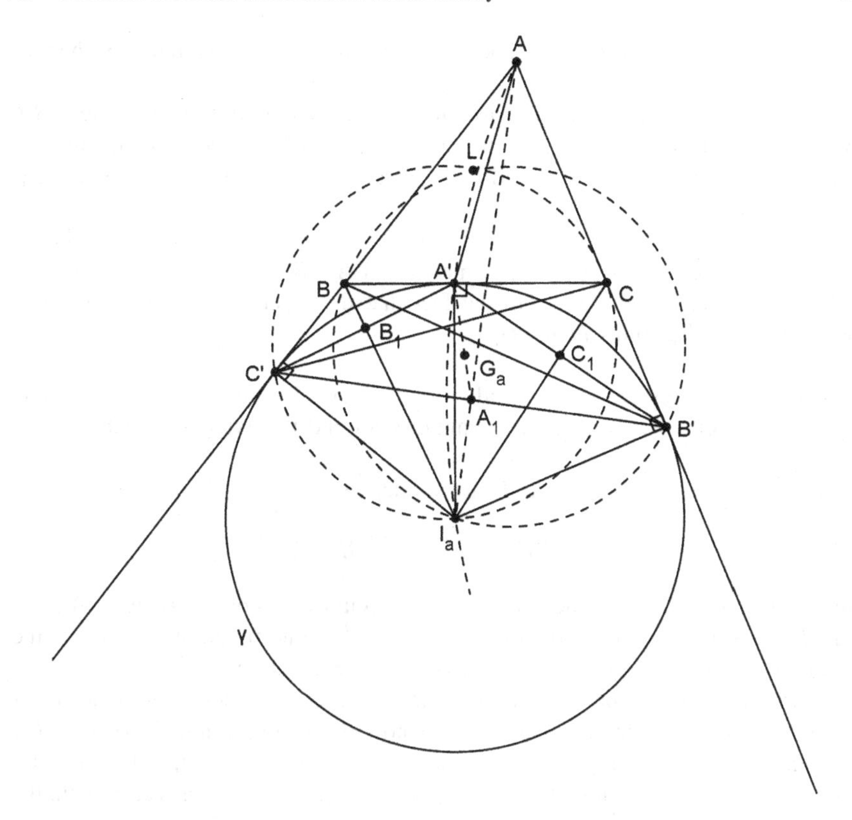

Fig. 6.36 Illustration of Problem 6.2.15

Therefore,

$$\sum_{\text{cyclic}} B_1C_1 \geq \sum_{\text{cyclic}} a \sin \frac{A}{2} \geq s. \tag{6.164}$$

□

6.2.15 Let I_a be the excenter corresponding to the side BC of a triangle ABC. Let A', B', C' be the tangency points of the excircle of center I_a with the sides BC, CA, and AB, respectively. Prove that the circumcircles of the triangles AI_aA', BI_aB', CI_aC' have a common point, different from I_a, situated on the line G_aI_a, where G_a is the centroid of the triangle $A'B'C'$.

(*Proposed by Dorin Andrica [20], Romania*)

Solution (by Michel Bataille, France) Let γ be the excircle of the triangle ABC corresponding to the side BC. Since $I_aA' = I_aC'$ and $BA' = BC'$, the line I_aB is the perpendicular bisector of $A'C'$ and intersects $A'C'$ at its midpoint B_1 (see Fig. 6.36).

Since $A'C'$ is the polar of B with respect to γ, the inversion in the circle γ exchanges B_1 and B.

Since B' is invariant under this inversion, the circumcircle of the triangle I_aBB' inverts into the median $B'B_1$ of the triangle $A'B'C'$. It also holds true that the circumcircles of the triangles I_aAA' and I_aCC' invert into the medians $A'A_1$ and $C'C_1$, respectively.

Therefore, all three circumcircles pass through the point I_a and through the inverse of G_a since G_a lies on the three medians $A'A_1$, $B'B_1$, and $C'C_1$.

The second result follows from the fact that the inverse of G_a is on the line passing through the points I_a and G_a.

Second solution Let E' be the midpoint of $C'A'$. We have $BA' \perp I_aA'$, while $BI_a \perp A'C'$ where $A'E' = C'E'$ by symmetry around the external bisector of angle $\widehat{B}$.

Thus, the triangles $BE'A'$, $A'E'I_a$ are similar. Hence

$$A'E' \cdot C'E' = \left(A'E'\right)^2 = BE' \cdot I_aE',$$

and the median $B'E'$ is the radical axis of the circumcircles of the triangles $A'B'C'$ and BI_aB'. Similarly, the median $C'F'$, where F' is the midpoint of $A'B'$, is the radical axis of the circumcircles of the triangles $A'B'C'$ and CI_aC'.

The point G_a where the medians $A'D'$, $B'E'$ and $C'F'$ meet has the same power with respect to the four circumcircles. Let now the second point P where I_aG_a meets the circumcircle of AI_aA'. Since I_aG_a is the radical axis of the circumcircles of AI_aA' and BI_aB', because I_a, G_a have the same power with respect to both, the point P also has the same power with respect to both circles. However, since it is on the circumcircle of the triangle AI_aA', it is also on the circumcircle of BI_aB'. Similarly, it is also on the circumcircle of the triangle CI_aC'. This completes the proof. □

6.2.16 Let C_1, C_2, C_3 be concentric circles with center point P and radii $R_1 = 1$, $R_2 = 2$, and $R_3 = 3$, respectively. Consider a triangle ABC with $A \in C_1$, $B \in C_2$, and $C \in C_3$. Prove that

$$\max S_{ABC} < 5,$$

where $\max S_{ABC}$ denotes the greatest possible area attained by the triangle ABC.

(*Proposed by Roberto Bosch Carbera [35], Cuba*)

Solution (by Roberto Bosch Carbera) Let A, B be the points such that the area of the triangle ABC becomes maximum, h_c be the length of the altitude from C, and P_c the foot of the altitude from the point P onto the side AB. It follows that (see Fig. 6.37)

$$h_c \leq PC + PP_c.$$

The equality holds if and only if the points C, P, P_c are collinear with P inside the segment CP_c. Now, $PC = 3$ and PP_c is fixed for given points A, B, or the

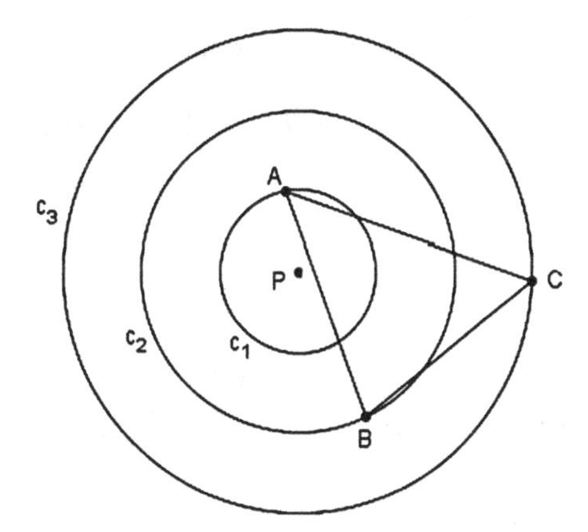

Fig. 6.37 Illustration of Problem 6.2.16

area becomes maximum when P is inside the segment CP_C, which is perpendicular to AB.

By cyclic symmetry, the point P is the orthocenter of the triangle ABC and it is inside the triangle ABC or the triangle ABC is acute.

It can be easily proved that if P is the orthocenter of an acute triangle ABC, we have

$$PA = 2R\cos\widehat{A}, \tag{6.165}$$

$$PB = 2R\cos\widehat{B}, \tag{6.166}$$

$$PC = 2R\cos\widehat{C}, \tag{6.167}$$

where R is the circumradius of the triangle ABC. Since $PC = 3$, by using (6.167), we get

$$3 = 2R\cos\widehat{C}$$

and thus

$$\begin{aligned} 3R &= 2R^2\cos\widehat{C} \\ &= 2R^2\sin\widehat{A}\sin\widehat{B} - 2R^2\cos\widehat{A}\cos\widehat{B} \\ &= \sqrt{4R^2-1}\sqrt{R^2-1} - 1, \end{aligned} \tag{6.168}$$

or

$$4R^3 - 14R - 6 = 0,$$

so

$$2R^3 = 7R + 3.$$

Additionally, using the formula

$$a^2b^2c^2 = \left(4R^2 - 1\right) \cdot \left(4R^2 - 4\right) \cdot \left(4R^2 - 9\right)$$
$$= 4\left(24R^3 + 49R^2 - 9\right), \tag{6.169}$$

we obtain

$$S^2 = \frac{a^2b^2c^2}{16R^2} = 6R + \frac{49}{4} - \frac{9}{4R^2}. \tag{6.170}$$

If $S \geq 5$, then

$$24R^3 + 49R^2 - 9 \geq 100R^2,$$

and taking into account that

$$2R^3 = 7R + 3,$$

we obtain

$$17R^2 - 28R - 9 \leq 0,$$

so

$$R \leq \frac{14 + \sqrt{349}}{17} < 2.$$

But if $R \leq 2$, then

$$4R^3 - 14R - 6 \leq 16R - 14R - 6 \leq -2, \tag{6.171}$$

which is a contradiction. Therefore, the area of the triangle ABC must be smaller than 5. □

6.2.17 Consider an angle $\widehat{xOy} = 60°$ and two points A, B moving on the sides Ox, Oy, respectively, so that $AB = a$, where a is a given straight line segment. Let AD, BE be the angle bisectors of $\widehat{A}$, $\widehat{B}$ in the triangle OAB. Determine the position for which the product

$$AE^m \cdot BD^n$$

attains its maximum value, when m, n are positive rational numbers expressing the lengths of two straight line segments.

Solution Analysis. Let us assume that this *maximizing* position does exist (see Fig. 6.38). We observe that

$$\widehat{EIA} = \widehat{BID} = \frac{\widehat{A}}{2} + \frac{\widehat{B}}{2} = \frac{180° - 60°}{2} = 60°. \tag{6.172}$$

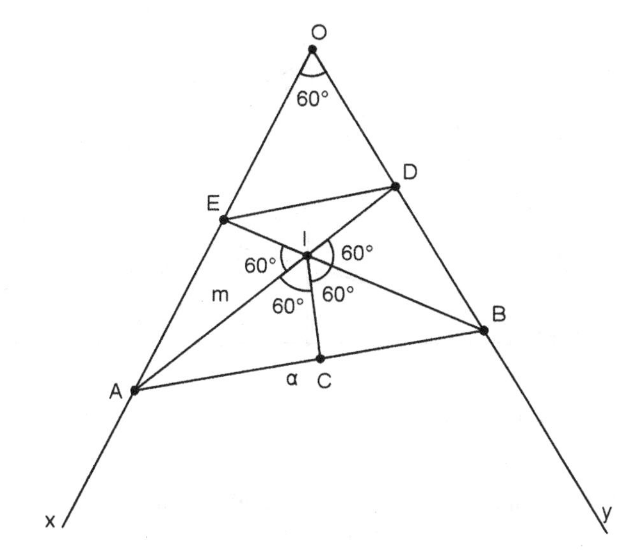

Fig. 6.38 Illustration of Problem 6.2.17

Consider the bisector IC of the angle $\widehat{AIB} = 180° - 60° = 120°$. We obtain

$$\widehat{EIA} = \widehat{AIC} = \widehat{CIB} = \widehat{BID} = 60°. \tag{6.173}$$

The triangles EAI and ACI are equal since AI is their common side and

$$\widehat{CAI} = \widehat{IAE}, \qquad \widehat{EIA} = \widehat{AIC} = 60°.$$

Thus

$$AE = AC.$$

In a similar manner, we prove that

$$BD = BC,$$

and thus

$$AE + BD = AC + CB = a, \tag{6.174}$$

where a is a constant.

It is a well known fact that, if for the positive real numbers

$$x_i > 0, \quad i = 1, \ldots, m$$

we have

$$\sum_{i=1}^{m} x_i = c,$$

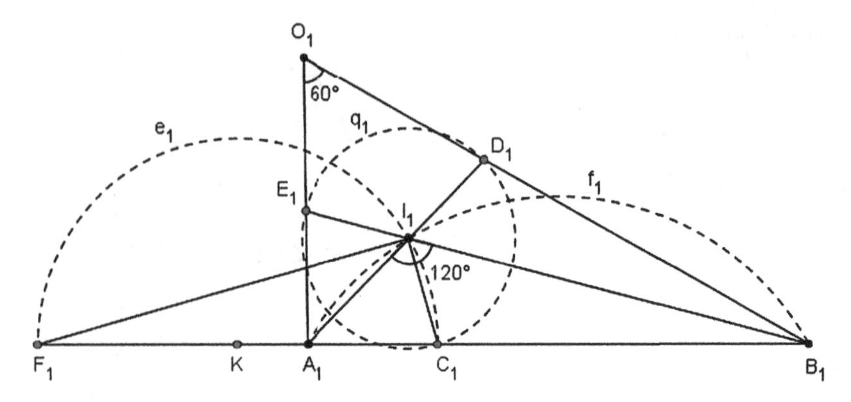

Fig. 6.39 Illustration of Problem 6.2.17

for a given constant $c \in \mathbb{R}^+$, then their product

$$\prod_{i=1}^{m} x_i^{\rho_i},$$

$\rho_i \in \mathbb{Q}$, attains its maximum value if

$$\frac{x_1}{\rho_1} = \frac{x_2}{\rho_2} = \cdots = \frac{x_m}{\rho_m} = \frac{c}{\rho_1 + \cdots + \rho_m}.$$

Therefore, in our case

$$\frac{AE}{m} = \frac{BD}{n} = \frac{a}{m+n} \quad \text{or} \quad \frac{AC}{m} = \frac{BD}{n} = \frac{a}{m+n}. \tag{6.175}$$

Construction–Synthesis. Consider a straight line segment $A_1B_1 = a$. We can determine points C_1, F_1 of the straight line segment A_1B_1 with C_1 in the interior of the line segment A_1B_1 and F_1 in its exterior such that (see Fig. 6.39)

$$\frac{A_1C_1}{C_1B_1} = \frac{F_1A_1}{F_1B_1} = \frac{m}{n} = \frac{I_1A_1}{I_1B_1}. \tag{6.176}$$

That is, the points C_1, F_1 are harmonic conjugates to the points A_1, B_1 with ratio $\frac{m}{n}$. Thus we can find the Apollonius circle e_1. We proceed by determining the arc f_1 such that its points *see* the line segment A_1B_1 under an angle of 120°. Let I_1 be the intersection of f_1 with e_1. Consider a circle with center I_1 and radius I_1C_1. Let D_1 be the intersection of this last circle with the straight line A_1E_1 and E_1 its intersection with the straight line B_1I_1.

Now, suppose that the intersection of the straight lines A_1E_1 and B_1D_1 is the point O_1. In this way, we take a point A belonging to the side Ox of the initial angle $\widehat{xOy}$, such that $OA = O_1A_1$ and on the side Oy we take a point B_1 so that $OB = O_1B_1$. This determines exactly the desired position. □

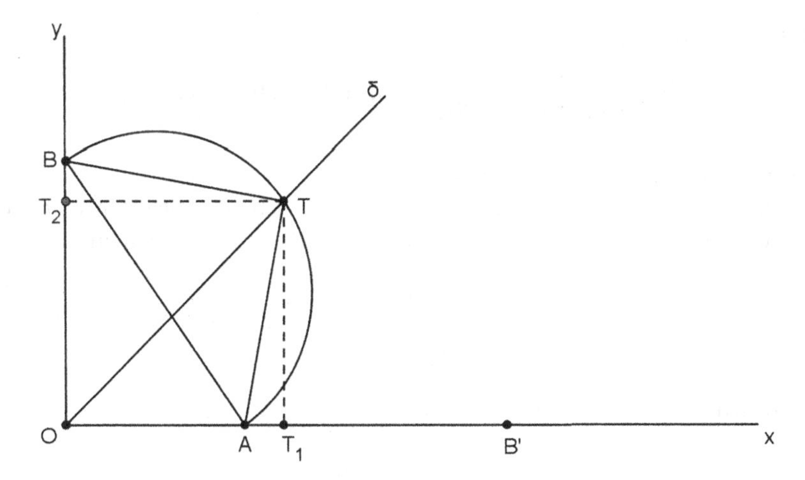

Fig. 6.40 Illustration of Problem 6.2.18

Note 2 Constructively, the same occurs on the other half plane that is determined by the straight line AB. However, because of the symmetry, the same result is recovered.

Hence, the position of the straight line segment AB can be constructed so that the product $AE^m \cdot BD^n$ attains its maximal value.

6.2.18 Let $\widehat{xOy} = 90°$ and points $A \in Ox$, $B \in Oy$ (with $A \neq O$, $B \neq O$), so that the condition

$$OA + OB = 2\lambda$$

holds, where $\lambda > 0$ is a given positive number. Prove that there exists a unique point $T \neq O$ such that

$$S_{OATB} = \lambda^2, \tag{6.177}$$

independently of the position of the straight line segment AB.

Solution On the straight line Ox, we choose a point B' so that $AB' = OB$. Let $O\delta$ be the bisector of the angle $\widehat{xOy}$. Consider the circle circumscribed to the triangle OAB intersecting the bisector at the point T (see Fig. 6.40).

It is evident that $\widehat{TAT_1} = \widehat{TBT_2}$ (since the quadrilateral $OATB$ is inscribed) and $TA = TB$ because $T \in O\delta$ where T_1 and T_2 are the projections of T onto Ox, Oy, respectively. It follows that the triangles OTB and ATB' are equal.

As a consequence, the equality $TO = TB'$ should hold and the point T should belong to the perpendicular bisector of $OB' = 2\lambda$ which, in its turn, is constant. Thus, we obtain that T is the intersection of the bisector $O\delta$ with the perpendicular bisector OB'. Consequently, the square OT_1TT_2 has been constructed with side length λ

and this actually means that

$$S_{OATB} = S_{OAT} + S_{OBT} = \frac{\lambda \cdot (OA + OB)}{2} = \lambda^2. \tag{6.178}$$

Thus the existence of the point T has been proved.

As far as the uniqueness of the existence is concerned, we proceed by contradiction. Suppose that there exists another point $T' \neq T$ such that the conditions

$$OA + OB = 2\lambda,$$

$$S_{T'OAB} = \lambda^2$$

are satisfied. Then

$$S_{TAB} = S_{T'AB},$$

and thus

$$TT' \parallel AB.$$

Hence, the last parallelism condition must be valid for any choice of position for the line segment AB. This leads to a contradiction. □

6.2.19 Let a given quadrilateral $A'B'C'D'$ be inscribed in a circle (O, R). Consider a straight line y intersecting the straight lines $A'D'$, $B'C'$, $B'A'$, and $D'C'$, at the points A, A_1, B, B_1, respectively, and also the circle (O, R) at the points M, M_1. Prove that

$$\sqrt{MA \cdot MA_1 \cdot MB \cdot MB_1} + \sqrt{M_1A \cdot M_1A_1 \cdot M_1B \cdot M_1B_1}$$
$$= \sqrt{(MA \cdot MA_1 + M_1A \cdot M_1A_1) \cdot (MB \cdot MB_1 + M_1B \cdot M_1B_1)}. \tag{6.179}$$

Solution According to the Cauchy–Schwarz–Buniakowski inequality, we have

$$\left(\sum_{i=1}^{n} x_i^2\right) \cdot \left(\sum_{i=1}^{n} y_i^2\right) \geq \left(\sum_{i=1}^{n} x_i \cdot y_i\right)^2. \tag{6.180}$$

The equality occurs if and only if

$$\frac{x_1}{y_1} = \frac{x_2}{y_2} = \cdots = \frac{x_n}{y_n}.$$

Applying inequality (6.180) in (6.179) for the case of equality, it should be enough to prove that

$$\frac{MA \cdot MA_1}{MB \cdot MB_1} = \frac{M_1A \cdot M_1B}{M_1B \cdot M_1B_1}. \tag{6.181}$$

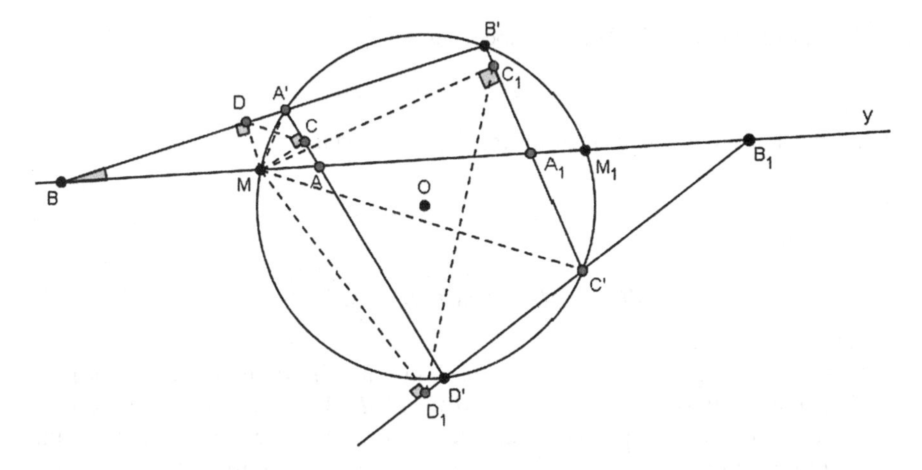

Fig. 6.41 Illustration of Problem 6.2.19

Consider $MD \perp A'B'$, $MC \perp A'D'$, $MC_1 \perp B'C'$, $MD_1 \perp D'C'$ (Fig. 6.41). In order to prove the equality (6.181), it is enough to verify that for a given direction of the straight line y, the ratio

$$\frac{MA \cdot MA_1}{MB \cdot MB_1}$$

is constant for any position of the line y. Let

$$\begin{aligned} MA &= k_1 \cdot MC, & MB &= k_2 \cdot MD, \\ MA_1 &= l_1 \cdot MC_1, & MB_1 &= l_2 \cdot MD_1, \end{aligned} \tag{6.182}$$

where k_1, k_2, l_1, l_2 are constant positive numbers and the ratio

$$\frac{MA \cdot MA_1}{MB \cdot MB_1}$$

becomes

$$\frac{MA \cdot MA_1}{MB \cdot MB_1} = \frac{k_1 k_2}{l_1 l_2} \cdot \frac{MC}{MD} \cdot \frac{MC_1}{MD_1} = \frac{k_1 k_2}{l_1 l_2}, \tag{6.183}$$

where the term

$$\frac{MC}{MD} \cdot \frac{MC_1}{MD_1} = 1,$$

since the triangles MCD, MC_1D_1 are similar and

$$\frac{MC}{MD} = \frac{MD_1}{MC_1}.$$

Indeed, the similarity of the triangles MCD, MC_1D_1 can be verified as follows.

The quadrilateral $MCA'D$ is inscribed, since

$$\widehat{D}+\widehat{C}=90^\circ+90^\circ=180^\circ.$$

The same holds true for the quadrilateral $MD_1C'C$ because

$$\widehat{D}_1+\widehat{C}_1=180^\circ.$$

Furthermore, the relations

$$\widehat{MDC}=\widehat{MA'C}=\widehat{MC'D'}=\widehat{MC_1D_1}$$

hold true.

The relation (6.183) is derived by observing that the straight line y preserves its direction and thus the right triangles MAC, MBD, MA_1C_1, and MB_1D_1 do preserve their angles for this particular direction. Hence, each one remains similar with respect to itself or, equivalently, they are representatives of cosine of constant angles. □

6.2.20 Let ABC be a triangle with $\widehat{BCA}=90^\circ$ and let D be the foot of the altitude from the vertex C. Let X be a point in the interior of the segment CD. Let K be the point on the segment AX, such that $BK=BC$. Similarly, let L be the point on the segment BX such that $AL=AC$. Let M be the point of intersection of AL and BK. Show that $MK=ML$.

(*53rd IMO, 2012, Mar del Plata, Argentina*)

Solution Consider the circles $C_1(B,BC)$, $C_2(A,AC)$, and $C(F,FK)$, where the circle C has its center on BK and it is internally tangential to the other two circles C_1, C_2 at K and L_1, respectively. The radial axes of the three circles will be intersected at a point of altitude CD. Let T be that point (see Fig. 6.42). We have $X\equiv CD\cap AK$.

Let the circle (T,TK), where $TK=TL_1$, intersects C_2 at a point I. The point B belongs to the radial axis L_1I, since the triangle is a right one at the vertex C and $BC=BK$.

Similarly, the straight line AK is the radial axis of the circles C_1, (T,TK). Thus, because of the uniqueness of the points, we deduce that

$$L_1\equiv L \quad\Rightarrow\quad F\equiv M.$$

This completes the proof. □

Remark 6.4 The point X is the radial center of the three circles C_1, C_2 and (T,TK).

6.2.21 Let AB be a straight line segment and C be a point in its interior. Let $C_1(D,r)$, $C_2(K,R)$ be two circles passing through A, B and intersecting each other orthogonally. If the straight line DC intersects the circle C_2 at the point M compute the supremum of $x\in\mathbb{R}$, where

$$x=S_{MAC}$$

denotes the area of the triangle MAC.

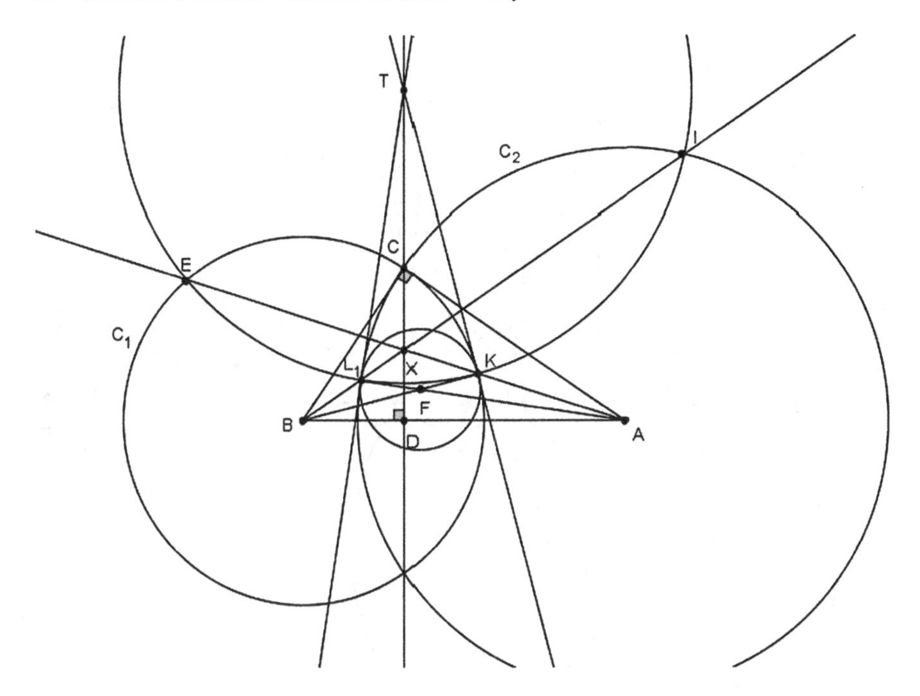

Fig. 6.42 Illustration of Problem 6.2.20

Solution Since the two circles C_1, C_2 are orthogonal, it follows that (see Fig. 6.43)

$$\widehat{KAD} = \widehat{DBK} = 90°.$$

It is a well known fact that the angle inscribed in a circle is equal to the angle formed by the corresponding chord and the tangent line at the edge of the chord. Thus

$$\widehat{MBA} = \widehat{MAD}.$$

Similarly, we get that $\widehat{BAM} = \widehat{DBM}$. Let the points A', B', be the projections of A, B, respectively on the straight line DC, then

$$\frac{S_{MBC}}{S_{MDA}} = \frac{BB' \cdot MC}{AA' \cdot DM} = \frac{BM \cdot BC}{AM \cdot AD}. \tag{6.184}$$

This is true because of the fact that the ratio of the areas of two triangles having one angle in common is equal to the fraction of the product of the sides that contain this angle. Similarly, we obtain

$$\frac{S_{MBD}}{S_{MAC}} = \frac{BB' \cdot DM}{AA' \cdot MC} = \frac{BM \cdot BD}{AC \cdot AM}. \tag{6.185}$$

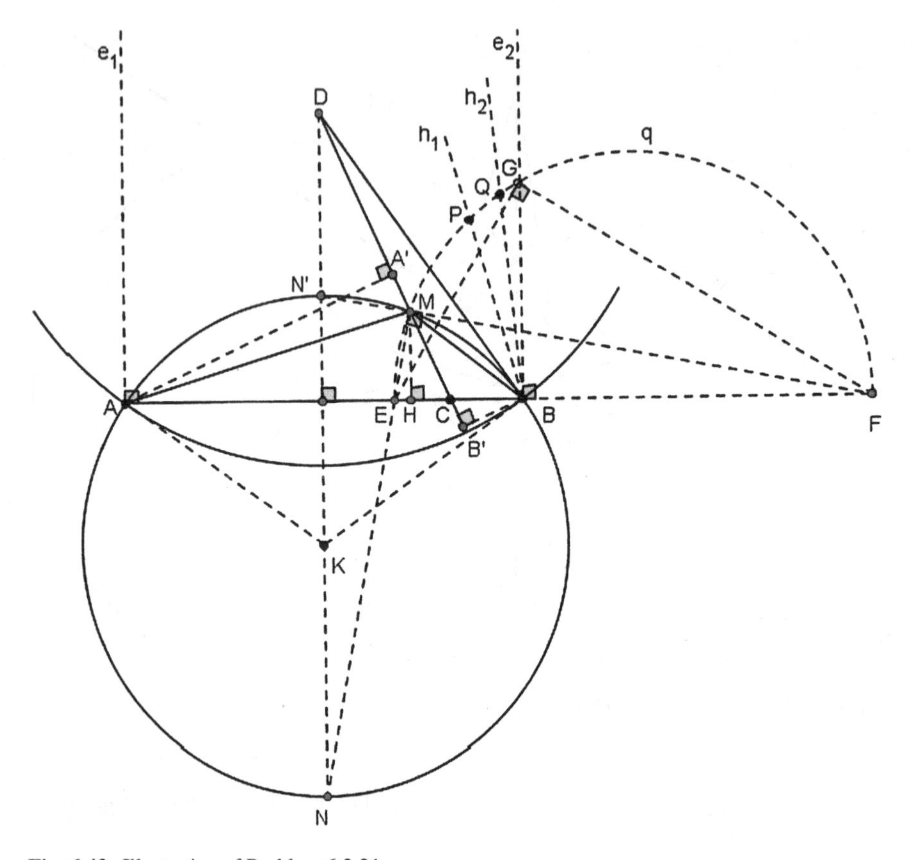

Fig. 6.43 Illustration of Problem 6.2.21

By (6.184) and (6.185), we obtain

$$\frac{B'B^2}{A'A^2} = \frac{BC}{AC} \cdot \left(\frac{BM}{AM}\right)^2, \tag{6.186}$$

where

$$\frac{BB'}{AA'} = \frac{BC}{AC}$$

(since the right triangles ACA', BCB' are similar to each other), and consequently,

$$\frac{MB^2}{MA^2} = \frac{BC}{AC}.$$

Thus

$$\frac{MB}{MA} = \sqrt{\frac{BC}{AC}}. \tag{6.187}$$

By (6.184), we obtain that the geometric locus of the point M is the arc GE (in the anticlockwise direction) of a circumference q with diameter EF, where G is the point of intersection of the perpendicular at the point B and of the straight line AB with the circle q. Additionally, it is symmetric with respect to the straight line AB, where E is a point in the interior of AB and F (in the exterior of AB, are the feet of the inner and of the outer bisector of the angle $\widehat{AMB}$, respectively, on the line AB) are harmonic conjugates of the points A, B with ratio $\sqrt{\frac{BC}{AC}}$, that is,

$$\frac{MB}{MA}=\frac{BE}{EA}=\frac{FB}{FA}=\sqrt{\frac{BC}{AC}}=\frac{m}{n}, \tag{6.188}$$

with

$$m=\sqrt{BC}, \qquad n=\sqrt{AC}.$$

It follows that

$$GB^2=BE\cdot BF.$$

Hence

$$GB=\sqrt{BE\cdot BF}, \tag{6.189}$$

since the angle $\widehat{EGB}=90^\circ$, where H is the foot of the projection of the point M on the straight line AB. However, it is a known fact that

$$EB=\frac{m\cdot BA}{m+n}, \qquad FB=\frac{m\cdot BA}{|m-n|}. \tag{6.190}$$

Hence

$$GB=\frac{m\cdot BA}{\sqrt{|m^2-n^2|}}, \tag{6.191}$$

and therefore

$$x<\frac{AB^2\cdot m}{2\sqrt{|m^2-n^2|}}. \tag{6.192}$$

The justification that the supremum is given by the quantity

$$\frac{AB^2\cdot m}{2\sqrt{|m^2-n^2|}}$$

yields from the following reasoning.

The set of points of the arc (G, E) of the semicircumference q, generated by the anticlockwise *motion* of the point G on the circle, has the following property:

For any point X of the arc (G, E), there exists a point T of (G, X) such that

$$\widehat{TBA} < 90^\circ.$$

Simultaneously, when the point M tends to coincide with the point G, then the straight line DA tends to coincide with the straight line $\epsilon_1 \perp AB$ and the straight line DA tends to coincide with the straight line $\epsilon_2 \perp AB$. This is actually the case when the triangle DAB degenerates. □

6.2.22 Let $ABCWD$ be a pentagon inscribed in a circle of center O. Suppose that the center O is located in the common part of the triangles ACD and BCW, where the point W is the intersection of the height of the triangle ACD, passing through the vertex A, with the circle. Let E be the intersection point of the straight line OK with the straight line AW, where K is the midpoint of the side AD. Suppose that the diagonal BW passes through the point E. Let Q be the common point of the diagonal BW with the straight line OK, such that $ZQ \parallel AW$ and Z be the point of intersection of the diagonals AC and BW. Compute the sum

$$\widehat{CDB} + \widehat{CBA}.$$

Solution We start by proving the following:

Lemma 6.8 *Let ABC be an acute triangle inscribed in a circle (O, R). Consider its heights AD, BF and its orthocenter H. Let*

$$E \equiv AD \cap (O, R), \qquad K \equiv BF \cap (O, R).$$

Then, there exists a point $N \neq K$ which belongs to the minor arc AC such that

$$LM = MN \quad \text{with } M \equiv BN \cap AC \text{ and } L \equiv BN \cap AE$$

if and only if

$$\widehat{EAN} = 90^\circ.$$

Proof of the Lemma Assume S is a point of the chord AN such that $HS \parallel BN$ and $P \equiv HS \cap AM$. We have $HP = PS$. By using the congruence theorem, and also since $HF = FK$ (this is true because the symmetrical points of the orthocenter of a triangle with respect to its sides are points on the circumscribed circle of the triangle) (see Fig. 6.44), we get

$$FP \parallel KS, \qquad \widehat{HKS} = 90^\circ \quad \text{and} \quad \widehat{ASH} = \widehat{ANB} = \widehat{AKB}, \tag{6.193}$$

which implies that the quadrilateral $AHSK$ can be inscribed in a circle and thus

$$\widehat{HAS} = 90^\circ.$$

Fig. 6.44 Lemma 6.8

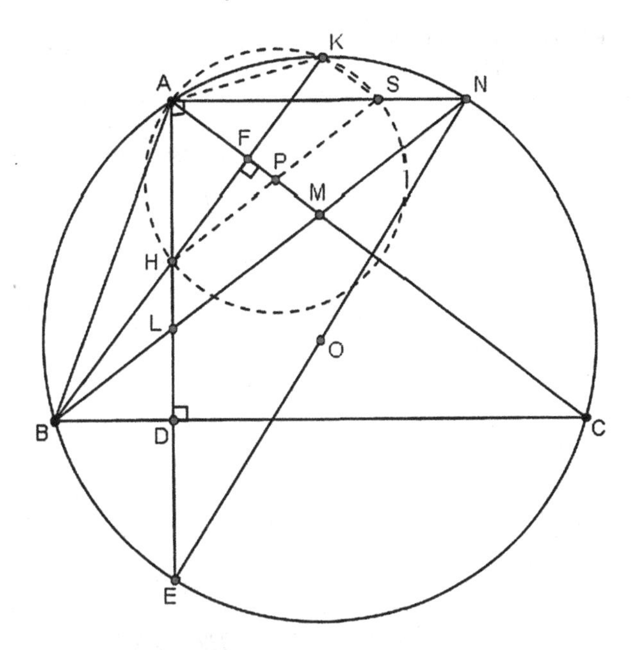

Consider now the point N to be the antidiametrical of E. Then $\widehat{HAN} = 90°$. We observe that the parallel to AC passing through the point K meets the straight line AN at the point S. Hence

$$\widehat{ASH} = \widehat{AKH} = \widehat{ANB},$$

and therefore,

$$HS \parallel LN \quad \text{implies } LM = MN.$$

This proves the assertion of Lemma 6.8. □

We have (see Fig. 6.45)

$$\widehat{EAC} = \widehat{ECA} = \widehat{EHD} = \widehat{EDH},$$

$$\widehat{QZE} = \widehat{ZEA} = 2\widehat{EAC} = 2\widehat{EHD},$$

therefore

$$\widehat{ZHQ} = \widehat{ZQH}$$

and

$$\widehat{HQE} = 90°.$$

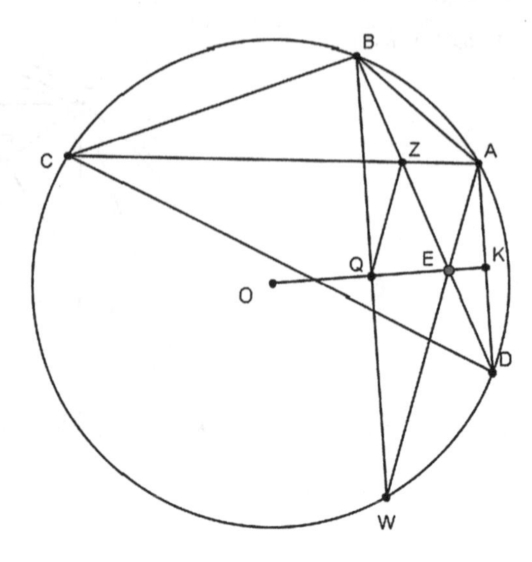

Fig. 6.45 Illustration of Problem 6.2.22

Hence

$$HZ = ZE.$$

By applying (6.193), it follows that the point E has to be the common point of the heights of the triangle. In this case, the triangle has to be isosceles with $BA = BC$. Hence

$$\widehat{CDB} + \widehat{CBA} = 180^\circ - \widehat{ADC} + \widehat{ADC} = 180^\circ. \tag{6.194}$$

□

6.2.23 On the straight line ϵ consider the collinear points A, B, C and let $AB > BC$. Construct the semicircumferences (O_1), (O_2) with diameters AB, BC, respectively, and let D, E be their intersection points with the semicircle (O) having as diameter the line segment O_1O_2. Define the points

$$D' \equiv (O_1) \cap DE, \qquad E' \equiv (O_2) \cap DE.$$

Prove that the points

$$P \equiv AD' \cap CE', \qquad Q \equiv AD \cap CE$$

and the midpoint M of the straight line segment AC are collinear.

(*Proposed by Kostas Vittas, Greece*)

Solution (by S.E. Louridas) The quadrilateral $BEQD$ is inscribed in a circle of diameter QB and center K (see Fig. 6.46). Consider

$$QQ' \perp DE, \qquad KK' \perp ED, \qquad BB' \perp ED.$$

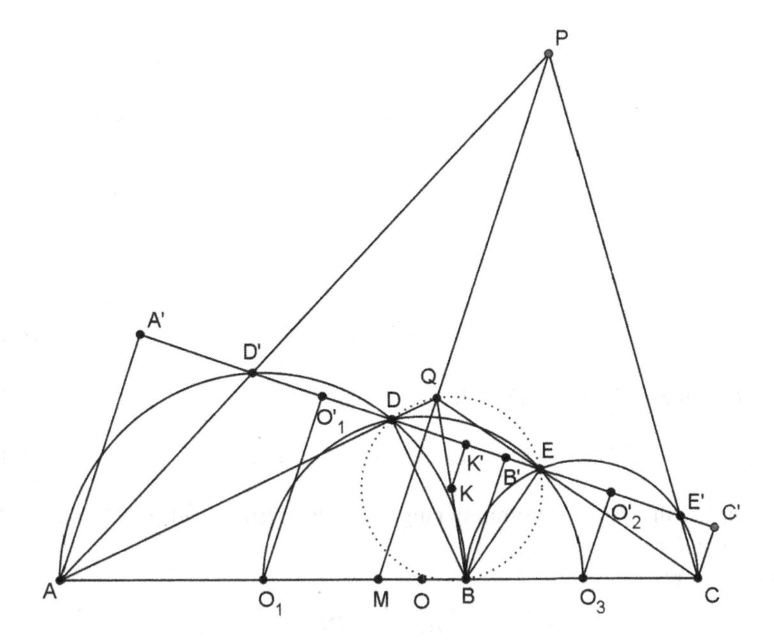

Fig. 6.46 Illustration of Problem 6.2.23

Therefore,

$$DK' = K'E \quad \text{and} \quad DQ' = EB'. \tag{6.195}$$

If $CC' \perp DE$, then from the trapezoid $BB'C'C$, we deduce

$$O_2O_2' \perp DE,$$

and thus

$$B'E = E'C' = Q'D. \tag{6.196}$$

Similarly, we get

$$AA' \perp DE,$$

and hence

$$A'D' = Q'E.$$

Thus

$$A'Q' = Q'C'. \tag{6.197}$$

Therefore, the straight line $Q'Q$ passes through the midpoint M of the straight line segment AC with $\widehat{QMB} = \widehat{BK'E}$. This is true since form the trapezoid $O_1O_1'O_2'O_2$

we have

$$O_1O_1'\perp DE,$$

hence

$$O_1'D = EO_2',$$

and thus

$$DD' = EE' \tag{6.198}$$

(where O is the midpoint of DE). It follows that

$$K'E \cdot K'E' = K'D \cdot K'D', \tag{6.199}$$

which implies that BK' is a common tangent (*radical axis*). Thus, the triangles PAC, BDE are similar.

Furthermore,

$$\widehat{BED} = \widehat{PCA}, \qquad \widehat{BDE} = \widehat{PAB}.$$

Since PM is a median of the triangle PAC, it follows that

$$\widehat{PMC} = \widehat{BK'E},$$

with

$$\widehat{QMB} = \widehat{BK'E},$$

from which we get

$$\widehat{QMB} = \widehat{PMC}.$$

The linear segment QM is perpendicular to DE, $K'B$ is perpendicular to MB and thus an inscribed quadrilateral is obtained. It follows that the point Q belongs to the straight line PM.

The straight line segment BK' is a median of the triangle $BE'D'$ which is similar to the triangle QAC with QM being its median. Hence

$$\widehat{QMC} = \widehat{BK'E}. \qquad \square$$

6.2.24 Let $\widehat{xOy}$ be an angle and A, B points in the interior of $\widehat{xOy}$. Investigate the problem of the constructibility of a point $C \in Ox$ such that

$$OD \cdot OE = OC^2 - CD^2, \tag{6.200}$$

where

$$D \equiv CA \cap Oy \quad \text{and} \quad E \equiv CB \cap Oy.$$

Solution We observe that for (6.200) to be valid, it should hold

$$OC > CD \quad \Rightarrow \quad \widehat{COD} < 90^\circ. \tag{6.201}$$

With no loss of generality, we can assume that $OA < OB$.

Suppose that such a point C does exist. Consider the circle with center C and radius CD intersecting the straight semiline Oy at the point E'. Then

$$OD \cdot OE = OC^2 - CD^2,$$

and thus

$$E \equiv E'. \tag{6.202}$$

Therefore, what we are looking for is a point $C \in Ox$ such that the triangle CDE is isosceles. Subject to the above, we are trying to find a way to apply the *power of a point* method.

Let $a = \widehat{xOy}$ and A' be the symmetric of A with respect to Ox. Hence

$$\widehat{CDE} = a + \widehat{OCD} = \frac{\pi - \widehat{DCE}}{2}.$$

Thus

$$2a + 2\widehat{OCD} = \pi - \widehat{DCE},$$

and therefore

$$2\widehat{OCD} + \widehat{DCE} = \pi - 2a. \tag{6.203}$$

Hence, the point C belongs to a constant arc, due to the fact that the points A' and B are fixed and

$$\widehat{A'CB} = \pi - 2a.$$

It follows that the point C is constructible in the case

$$\widehat{xOy} < \frac{\pi}{2}.$$

□

6.2.25 Let ABC be a triangle satisfying the following property: There exists an interior point L such that

$$\widehat{LBA} = \widehat{LCA} = 2\widehat{B} + 2\widehat{C} - 270^\circ.$$

Let B', C' be the symmetric of the points B and C with respect to the straight lines AC and AB, respectively. Prove that

$$AL \perp C'B'.$$

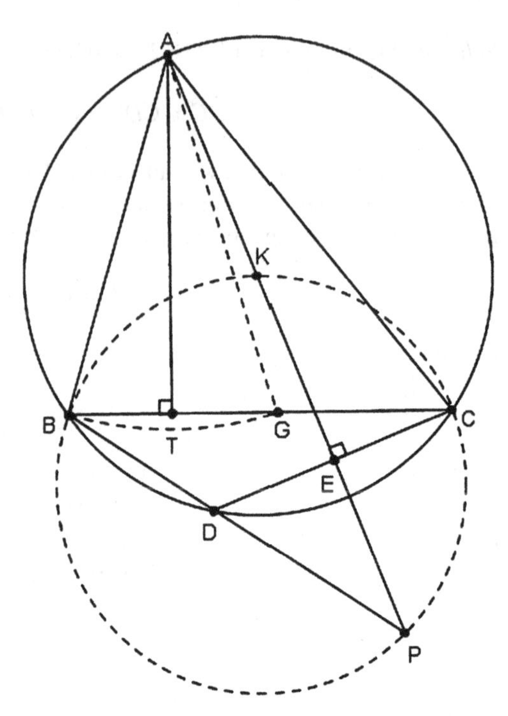

Fig. 6.47 Lemma 6.9

Solution We shall make use of the following auxiliary lemma:

Lemma 6.9 *Let ABC be a triangle and K be the center of the circumscribed circle γ. Let P be a point of the arc BC of a circle which contains K. If*

$$\widehat{PBC} = \pi - 2\widehat{B} \quad \text{and} \quad \widehat{PCB} = \pi - 2\widehat{C},$$

then

$$\widehat{BPC} = \pi - 2\widehat{A}.$$

Let T be the foot T of the altitude AT. Then

$$BP - PC = TC - TB. \tag{6.204}$$

Proof of the Lemma We base our study on Fig. 6.47. We start by pointing out that the points K, B, P, C are concyclic. This fact leads to the conclusion that the points A, K, P are collinear. Indeed,

$$\widehat{PKC} = \widehat{PBC} = \pi - 2\widehat{B},$$

and thus

$$\widehat{PKC} = 2\widehat{KCA}. \tag{6.205}$$

Therefore, the collinearity follows. Let us assume that

$$G \in BC, \qquad BT = TG.$$

Then we obtain

$$\widehat{AGC} = \widehat{ABP} \quad \text{and} \quad AG = AB.$$

Additionally, assume that D belongs to the arc BP and $PC = PD$.
Then

$$\widehat{PDC} = \widehat{A},$$

and thus the points A, B, C, D are concyclic. Hence

$$\widehat{ADB} = \widehat{C}$$

and

$$BP - PC = BP - PD = BD.$$

From the equality of the triangles ABD, AGC, we deduce that the assertion of the lemma holds true since the equality yields $BD = GC$.

Going back to the main problem, let AT be the height of our triangle and K the center of its circumscribed circle. Then S has to be the center of the circle (KBC). It is enough to prove that

$$(AC')^2 - (AB')^2 = (LC')^2 - (LB')^2.$$

Equivalently, it is enough to prove that

$$(AC)^2 - (AB)^2 = [(LC')^2 - R^2] - [(LB')^2 - R^2].$$

However,

$$\begin{aligned} TC^2 - BC^2 &= BC(BC + CP) - BC(BC + CP) \\ &= BC \cdot BP - BC \cdot CP, \end{aligned}$$

and thus

$$BP - PC = TC - TB.$$

The above relation is true in virtue of Lemma 6.9 (see Fig. 6.48). This completes the proof. □

6.2.26 Let $AB = a$ be a straight line segment. On its extension towards the point B, consider a point C such that $BC = b$. With diameter the straight line segments AB and AC we construct two semicircumferences on the same side of the straight line AC. The perpendicular bisector to the straight line segment BC intersects the exterior semicircumference at a point E. Prove or disprove the following assertion 1 and solve problem 2:

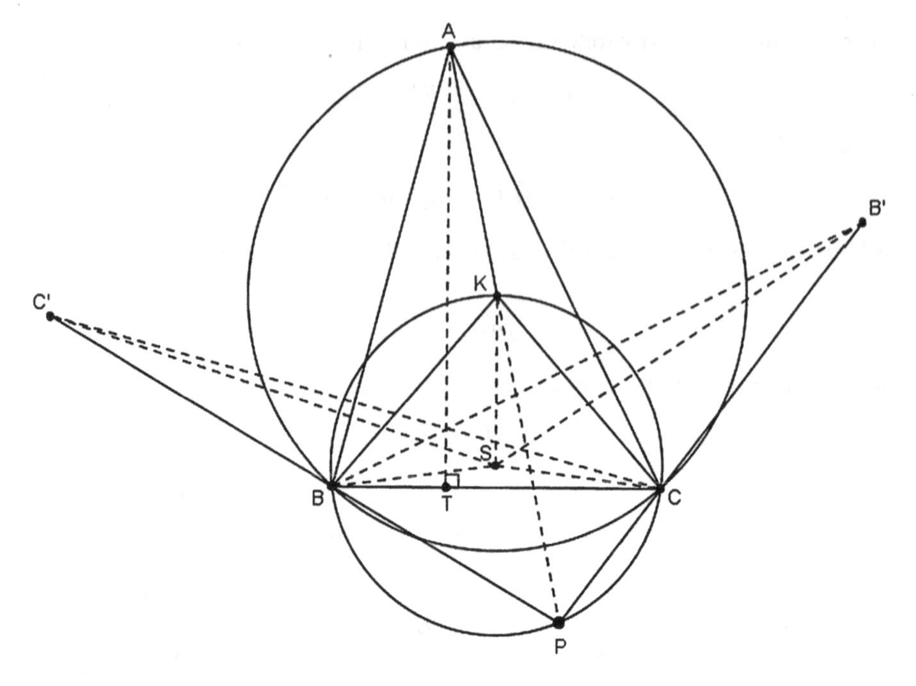

Fig. 6.48 Illustration of Problem 6.2.25

1. There exists a circle inscribed in the curved triangle $ABEA$.
2. If K is the center of the previously inscribed circle and M is the point of intersection of the straight line BK with the semicircumference of diameter AC, compute the area of the domain that is bounded from the semicircumference of diameter AC and the perimeter of the triangle MAC.

Solution Let us consider the perpendicular to AC passing through the point B and intersecting the semicircumference (D, R) of diameter AC at the point M (see Fig. 6.49). Let (O, ρ) be the circle of center belonging to the straight line segment BM, externally tangential to the semicircle of diameter AB and internally tangential to the semicircle (D, R). This is a classic Apollonius construction.

Furthermore, let us assume that (K, k) is the new circle (we are actually investigating its existence). Let BF be the tangent to the semicircle (K, k). We are going to prove that the triangles KBF and EHC are similar if H is the midpoint of BC and E is a point in the semicircle (D, R) with $EH \perp BC$. Indeed, the following relations hold true

$$KB^2 = (R-k)^2 - (2\rho - R)^2 = k^2 + 4\rho R - 2kR - 4\rho^2, \tag{6.206}$$

$$KB^2 = (\rho + k)^2 - \rho^2 = 2\rho k + k^2, \tag{6.207}$$

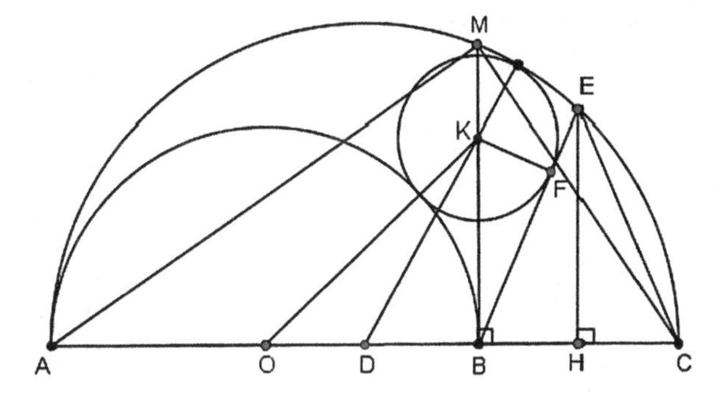

Fig. 6.49 Illustration of Problem 6.2.26

and thus

$$\left(\frac{KB}{k}\right)^2 = \frac{2\rho}{k} + 1. \tag{6.208}$$

The relations (6.206), (6.208) yield

$$k = \frac{2\rho(R-\rho)}{R+\rho},$$

and by (6.207), we obtain

$$\left(\frac{KB}{k}\right)^2 = \frac{2R}{R-\rho} \tag{6.209}$$

with

$$\left(\frac{EC}{CH}\right)^2 = \frac{2R \cdot CH}{CH^2} = \frac{2R}{R+\rho}. \tag{6.210}$$

Finally, from (6.209) and (6.210) we derive

$$\frac{KB}{KF} = \frac{EC}{CH},$$

thus

$$ECH \sim KBF,$$

and therefore the triangles

$$FKB, \quad FBH, \quad \text{and} \quad FCH$$

are equal. By the relation (6.208), we obtain that the straight line segment BF passes through the point E, and thus

$$MB^2 = ab \quad \Rightarrow \quad MB = \sqrt{ab}.$$

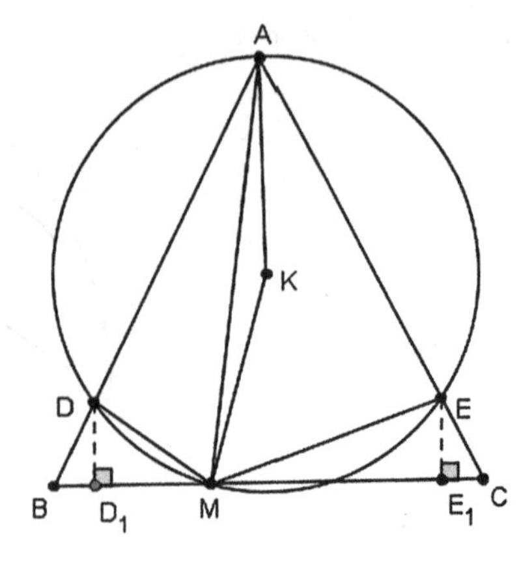

Fig. 6.50 Illustration of Problem 6.2.27

Therefore,

$$S_{MAC} = \frac{(a+b)\sqrt{ab}}{2}. \tag{6.211}$$

Hence for the area S we have

$$S = \frac{\pi(a+b)^2}{2} - \frac{(a+b)\sqrt{ab}}{2}. \tag{6.212}$$

□

6.2.27 Let ABC be a triangle with $AB \geq BC$. Consider the point M on the side BC and the isosceles triangle KAM with $KA = KM$. Let the angle $\widehat{AKM}$ be given such that the points K, B are in different sides of the straight line AM satisfying the condition

$$360^\circ - 2\widehat{B} > \widehat{AKM} > 2\widehat{C}.$$

The circle (K, KA) intersects the sides AB, AC at the points D and E, respectively. Find the position of the point $M \in BC$ so that the area of the quadrilateral $ADME$ attains its maximum value.

Solution Using the inequality

$$360^\circ - 2\widehat{B} > \widehat{AKM},$$

we get

$$\widehat{MDA} > \widehat{B}, \tag{6.213}$$

and similarly, using the inequality $\widehat{AEM} > \widehat{C}$, we deduce (see Fig. 6.50)

$$\widehat{BDM} > \widehat{A}. \tag{6.214}$$

The inequalities (6.213) and (6.214) guarantee that D, E are interior points of the sides AC and AB, respectively. The angle $\widehat{AKM}$ is by assumption of fixed measure, hence the inscribed quadrilateral (cyclic) $ADME$ has angles which preserve their measure and consequently the triangles DMB and EMC preserve the measure of their angles. The triangle ABC is fixed; therefore, the area of the quadrilateral $ADME$ attains its maximal value if and only if the sum S where

$$S = S_{BDM} + S_{CEM}$$

attains its minimal value.

We have

$$2S = BM \cdot DD_1 + MC \cdot EE_1,$$

where $DD_1 \perp BC$ and $EE_1 \perp BC$. Since the triangles BMD and EMC preserve their angles (they remain similar to themselves during the procedure) there exist positive constants k, l such that

$$DD_1 = k \cdot BM = k \cdot x$$

and

$$EE_1 = l \cdot MC = l \cdot y,$$

with $BM = x$ and $MC = y$. Hence,

$$2S = k \cdot x^2 + l \cdot y^2, \tag{6.215}$$

under the constraint $x + y = a$, where a denotes the length of the side BC. It follows that (6.215) assumes the form

$$(k+l)x^2 - 2alx + la^2 - 2S = 0. \tag{6.216}$$

Equation (6.216) admits a real solution if and only if

$$S \geq \frac{kla^2}{2(k+l)}, \tag{6.217}$$

and thus the minimum of the quantity $2S$ is achieved for

$$S_{\min} = \frac{kla^2}{2(k+l)}.$$

In this case, it holds

$$x = \frac{ak}{k+l}, \qquad y = \frac{al}{k+l}.$$

Therefore, the point $M \in BC$ is the point that divides the side BC in ratio l/k. □

6.2.28 Let $ABCD$ be a cyclic quadrilateral, $AC = e$ and $BD = f$. Let us denote by r_a, r_b, r_c, r_d the radii of the incircles of the triangles BCD, CDA, DAB, ABC, respectively. Prove the following equality

$$e \cdot r_a \cdot r_c = f \cdot r_b \cdot r_d. \tag{6.218}$$

(Proposed by Nicuşor Minculete and Cătălin Barbu, Romania)

Solution (by N. Minculete and C. Barbu) In any triangle ABC, we have

$$r = \frac{b + c - a}{2} \tan \frac{A}{2},$$

where a, b, c are the lengths of the sides BC, CA, AB and r is the inradius of the triangle ABC. We apply this relation to the triangles BCD and ABD, and we get

$$r_a = \frac{b + c - f}{2} \tan \frac{C}{2}, \qquad r_c = \frac{a + d - f}{2} \tan \frac{A}{2}.$$

But

$$\tan \frac{A}{2} \tan \frac{C}{2} = 1$$

because $A + C = \pi$. Therefore, we obtain

$$4r_a r_c = ab + cd + ac + bd - f(a + b + c + d) + f^2.$$

But, from Ptolemy's first theorem, we have

$$ef = ac + bd.$$

Thus, we obtain

$$4r_a r_c = ab + cd + f(e + f - a - b - c - d).$$

Multiplying by e, we obtain

$$4er_a r_c = e(ab + cd) + ef(e + f - a - b - c - d).$$

Similarly, we can deduce that

$$4fr_b r_d = f(ad + bc) + ef(e + f - a - b - c - d).$$

Combining the above relations with Ptolemy's second theorem, we obtain

$$\frac{e}{f} = \frac{ad + bc}{ab + cd},$$

from which the desired result follows. □

6.2.29 Prove that for any triangle the following equality holds

$$-\frac{a^2}{r}+\frac{b^2}{r_c}+\frac{c^2}{r_b}=4R-4r_a, \tag{6.219}$$

where a, b, c are the sides of the triangle, R is the radius of the circumscribed circle, r is the corresponding radius of the inscribed circle and r_a, r_b, r_c are the radii of the corresponding exscribed circles of the triangle.

(*Proposed by Nicuşor Minculete and Cătălin Barbu, Romania*)

Solution (by N. Minculete and C. Barbu) For any triangle, we have:

$$r_a=\frac{S}{s-a},\qquad r_b=\frac{S}{s-b},\qquad r_c=\frac{S}{s-c},\qquad r=\frac{S}{s},$$

$$abc=4RS,$$

$$b^2+c^2-a^2=2bc\cos A,$$

$$\sin\frac{A}{2}=\sqrt{\frac{(s-b)(s-c)}{bc}},$$

where S is the area of the triangle and s is the semiperimeter of the triangle.

It follows that

$$\begin{aligned}
-\frac{a^2}{r}+\frac{b^2}{r_c}+\frac{c^2}{r_b}&=\frac{1}{S}\big(-a^2s+b^2(s-c)+c^2(s-b)\big)\\
&=\frac{1}{S}\big(s\big(b^2+c^2-a^2\big)-bc(b+c)\big)\\
&=\frac{1}{S}(2sbc\cos A-2bcs+abc)\\
&=\frac{1}{S}\big(abc+2sbc(\cos A-1)\big)\\
&=\frac{1}{S}\left(4RS-2bcs\cdot 2\sin^2\frac{A}{2}\right)\\
&=\frac{1}{S}\left(4RS-\frac{4s(s-a)(s-b)(s-c)}{s-a}\right)\\
&=\frac{1}{S}(4RS-4Sr_a)=4R-4r_a,
\end{aligned}$$

and this completes the proof. □

6.2.30 For the triangle ABC let $(x, y)_{ABC}$ denote the straight line intersecting the union of the straight line segments AB and BC at the point X and the straight line segment AC at the point Y in such a way that the following relation holds

$$\frac{\widetilde{AX}}{AB+BC}=\frac{AY}{AC}=\frac{xAB+yBC}{(x+y)(AB+BC)},$$

where $\widetilde{AX}$ is either the length of the line segment AX, in case X lies between the points A, B, or the sum of the lengths of the straight line segments AB and BX if the point X lies between B and C. Prove that the three straight lines $(x, y)_{ABC}$, $(x, y)_{BCA}$, and $(x, y)_{CBA}$ concur at a point which divides the straight line segment NI in a ratio $x : y$, where N is the Nagel's point and I the incenter of the triangle ABC.

(*Proposed by Todor Yalamov, Sofia University, Bulgaria*)

Solution (by Peter Y. Woo, California, USA and extension by the editor of Crux Mathematicorum) As usual, we consider

$$a=BC, \qquad b=CA, \qquad c=AB,$$

and

$$s=\frac{a+b+c}{2}.$$

Let

$$t=\frac{x}{x+y},$$

so

$$1-t=\frac{y}{x+y},$$

and the ratio we are interested in becomes a function of t, that is,

$$f(t)=\frac{tc+(1-t)a}{a+c}=\frac{\tilde{AX}_t}{a+c}=\frac{AY_t}{b},$$

where X_t is the point of the union $AB \cup BC$ and Y_t is the point of AC that correspond to the parameter t. Based on the assumption, we have

$$(x, y)_{ABC}=X_tY_t.$$

Of course, if $a=c$, the function $f(t)$ is constant and $(x, y)_{ABC}$ is the straight line NI for all the x, y. This is compatible with what we want to prove, unless

$$a=b=c,$$

that is, when the triangle ABC is equilateral. In this case,

$$N \equiv I$$

(which implies that the straight line NI does not exist) and our three straight lines meet at this point. So, let us assume that $a \neq c$, then the function $f(t)$ is non-constant and the straight lines NI and $(x, y)_{ABC}$ intersect. We shall see that $(x, y)_{ABC}$

is a straight line of the family of the straight lines which are parallel to the angle bisecant of the angle $\hat{B}$ (where B is the middle vertex to the subtenant) and that divides the straight line segment to the ratio

$$\frac{t}{1-t}=\frac{x}{y}.$$

This is a consequence of the fact that $(x,y)_{ABC}$ and $(x,y)_{CBA}$ are representing the same straight line, so $(x,y)_{CBA}$ is the third straight line. For the problem that concerns us and under the hypothesis $0\le t\le 1$, we observe in particular:

$$AY_1=b\cdot f(1)=\frac{bc}{a+c},\qquad CY_1=b-AY_1=\frac{ab}{a+c},$$

where Y_1 is the foot of the angle bisecant of the angle $\widehat{CBA}$ (since this divides the side AC in ratio $c:a$),

$$AY_0=bf(0)=\frac{ab}{a+c},\qquad CY_0=\frac{bc}{a+c},$$

$$X_1=B,$$

since

$$f(1)=\frac{c}{a+c}=\frac{AX_1}{a+c},$$

$$A\tilde{X}_0=(a+c)f(0)=a,$$

and $c\ge a$, X_0 lies on the straight line AB (where $CX_0=c$ and $BX_0=c-a$) otherwise, when $a\ge c$ the point X_0 lies on the straight line segment BC (where $CX_0=c$ and $BX_0=a-c$).

It follows that the straight line X_1Y_1 bisects the angle $\widehat{CBA}$, therefore passes from the incenter I. We are going to prove that the straight line X_0Y_0 passes from N Nagel's point (see Fig. 6.51). In order to determine N we use the points P, Q, R where the exscribed circles are intersecting the sides BC, CA and AB of the triangle ABC, where

$$BR=CQ=s-a,\qquad AR=CP=s-b,\qquad AQ=BP=s-c.$$

The Nagel's point is defined as the common point of AP, BQ and CR. By applying Menelaus' theorem, with bisecant NCR, to the triangle BQA, we deduce that

$$\begin{aligned}\frac{BN}{NQ}&=\frac{CA}{QC}\cdot\frac{RB}{AR}\\&=\frac{b}{s-a}\cdot\frac{s-a}{s-b}=\frac{b}{s-b}.\end{aligned}\qquad(6.220)$$

Fig. 6.51 Illustration of Problem 6.2.30

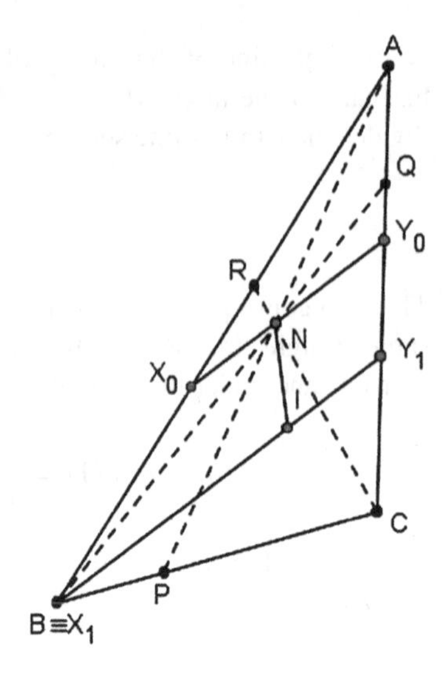

Let

$$N' = X_0Y_0 \cap BQ.$$

We want to show that $N' \equiv N$. At this point we will need the length

$$QY_0 = |AY_0 - AQ|$$
$$= \left|\frac{ab}{a+c} - (s-c)\right| = \frac{|a-c|(s-b)}{a+c}. \tag{6.221}$$

When $c > a$, we apply Menelaus' theorem with bisecant $N'Y_0X_0$ for the triangle BQA and obtain

$$\frac{BN'}{N'Q} = \frac{Y_0A}{QY_0} \cdot \frac{X_0B}{AX_0} = \frac{ab}{a+c} \cdot \frac{a+c}{(c-a)(s-b)} \cdot \frac{c-a}{a} = \frac{b}{s-b}. \tag{6.222}$$

If $a > c$, we apply Menelaus' theorem with bisecant $N'Y_0X_0$ for the triangle BCQ to get

$$\frac{BN'}{N'Q} = \frac{Y_0C}{QY_0} \cdot \frac{X_0B}{CX_0}$$
$$= \frac{bc}{a+c} \cdot \frac{a+c}{(a-c)(s-b)} \cdot \frac{a-c}{c} = \frac{b}{s-b}. \tag{6.223}$$

In both cases, the term $\frac{BN'}{N'Q}$ is equal to the value of the term $\frac{BN}{NQ}$ appearing in Eq. (6.220). From this fact we conclude that $N \equiv N'$, and X_0Y_0 intersects IN at the point N, as it is desired (see Fig. 6.52).

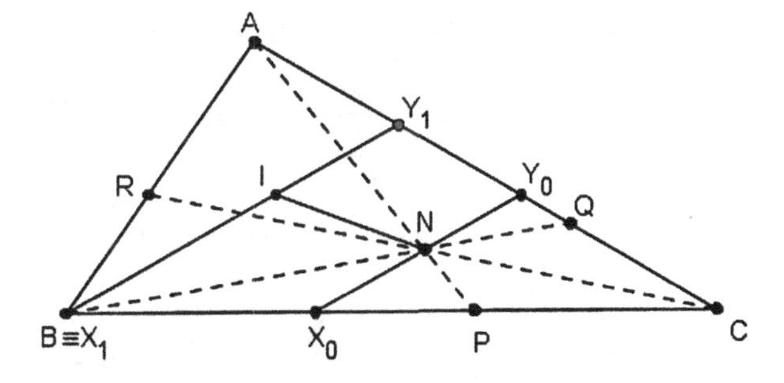

Fig. 6.52 Illustration of Problem 6.2.30

It remains to observe that $X_0Y_0 \parallel X_1Y_1$, where $c > a$ (this happens when X_0 lies on AB)

$$\frac{AX_0}{AX_1} = \frac{a}{c} = \frac{ab}{a+c} \cdot \frac{bc}{a+c} = \frac{AY_0}{AY_1}, \tag{6.224}$$

when $a > c$ (and X_0 lies on BC)

$$\frac{CX_0}{CX_1} = \frac{c}{a} = \frac{bc}{a+c} \cdot \frac{ab}{a+c} = \frac{CY_0}{CY_1}. \tag{6.225}$$

Finally, since X_t divides the line segment X_0X_1 in a fractional expression

$$\frac{t}{1-t},$$

and because of the fact that Y_t divides Y_0Y_1 in the same fraction, the line segment X_tY_t is parallel to both X_0Y_0 and X_1Y_1 for all t's. It follows that X_tY_t intersects the segment IN at the point which divides NI in the same ratio

$$\frac{t}{1-t} = \frac{x}{y}.$$

This completes the proof. □

6.2.31 Let T be the Torricelli's point of the convex polygon $A_1A_2 \dots A_n$ and (d) a straight line such that $T \in (d)$ and $A_k \notin (d)$, where $k = 1, 2, \dots, n$. If we denote by $B_1, B_2, \dots, B_n$ the projections of the vertices $A_1, A_2, \dots, A_n$ on the line (d), respectively, prove that

$$\sum_{k=1}^{n} \frac{\overrightarrow{TB_k}}{TA_k} = \overrightarrow{0}.$$

(*Proposed by Mihály Bencze, Braşov, Romania*)

Solution Let $d \equiv (0x)$, $T \equiv (0x) \cap (0y)$, $A_k \equiv (x_k, y_k)$ and $B_k \equiv (x_k, 0)$, where $k = 1, 2, \dots, n$.

If we denote by M the point $(x, 0)$, then we can write

$$f(x) = \sum_{k=1}^{n} MA_k = \sum_{k=1}^{n} \sqrt{(x - x_k)^2 + y_k^2}.$$

Since T is a Torricelli's point, it follows that $f(x)$ attains its minimal value at

$$f(0) = \sum_{k=1}^{n} \sqrt{x_k^2 + y_k^2} = \sum_{k=1}^{n} TA_k.$$

It is evident that the function $f : \mathbb{R} \to \mathbb{R}$ is continuous and differentiable.

We have

$$f(0) = \sum_{k=1}^{n} TA_k \le \sum_{k=1}^{n} MA_k = f(x),$$

for every $x \in \mathbb{R}$. Therefore, from Fermat's theorem we obtain $f'(0) = 0$, that is,

$$\sum_{k=1}^{n} \frac{x_k}{\sqrt{x_k^2 + y_k^2}} = 0.$$

But

$$x_k = \|\overrightarrow{TB_k}\| \quad \text{and} \quad TA_k = \sqrt{x_k^2 + y_k^2}.$$

Thus

$$\sum_{k=1}^{n} \frac{\overrightarrow{TB_k}}{TA_k} = \overrightarrow{0}.$$

□

6.2.32 Let $ABCD$ be a quadrilateral. We denote by E the midpoint of the side AB, F the centroid of the triangle ABC, K the centroid of the triangle BCD, and G the centroid of the given quadrilateral. For all points M of the plane of the quadrilateral, different from A, E, F, G, prove the following inequality

$$\frac{6MB}{MA \cdot ME} + \frac{2MC}{ME \cdot MF} + \frac{MD}{MF \cdot MG} \ge \frac{5MK}{MA \cdot MG}.$$

(Proposed by Mihály Bencze, Braşov, Romania)

Solution We have

$$\frac{z_2}{z_1(z_1 + z_2)} = \frac{1}{z_1} - \frac{1}{z_1 + z_2},$$

$$\frac{z_3}{(z_1 + z_2)(z_1 + z_2 + z_3)} = \frac{1}{z_1 + z_2} - \frac{1}{z_1 + z_2 + z_3},$$

$$\frac{z_4}{(z_1+z_2+z_3)(z_1+z_2+z_3+z_4)} = \frac{1}{z_1+z_2+z_3} - \frac{1}{z_1+z_2+z_3+z_4}.$$

Adding the above identities, we obtain

$$\frac{z_2}{z_1(z_1+z_2)} + \frac{z_3}{(z_1+z_2)(z_1+z_2+z_3)} + \frac{z_4}{(z_1+z_2+z_3)(z_1+z_2+z_3+z_4)} = \frac{z_2+z_3+z_4}{z_1(z_1+z_2+z_3+z_4)}.$$

It follows that

$$\frac{|z_2+z_3+z_4|}{|z_1||z_1+z_2+z_3+z_4|} \le \frac{|z_2|}{|z_1||z_1+z_3|} + \frac{|z_3|}{|z_1+z_2||z_1+z_2+z_3|} + \frac{|z_4|}{|z_1+z_2+z_3||z_1+z_2+z_3+z_4|}.$$

If $A(a)$, $B(b)$, $C(c)$, $D(d)$, $E((a+b)/2)$, $F((a+b+c)/3)$, $K((b+c+d)/3)$, $G((a+b+c+d)/4)$, $M(z)$, $z_1 = z-a$, $z_2 = z-b$, $z_3 = z-c$, and $z_4 = z-d$, then we get

$$\frac{3|z-\frac{b+c+d}{3}|}{4|z-a||z-\frac{a+b+c+d}{4}|} \le \frac{|z-b|}{2|z-a||z-\frac{a+b}{2}|} + \frac{|z-c|}{2|z-\frac{a+b}{2}|\cdot 3|z-\frac{a+b+c}{3}|} + \frac{|z-d|}{3|z-\frac{a+b+c}{3}|\cdot 4|z-\frac{a+b+c+d}{4}|},$$

or

$$\frac{6MB}{MA\cdot ME} + \frac{2MC}{ME\cdot MF} + \frac{MD}{MF\cdot MG} \ge \frac{5MK}{MA\cdot MG}.$$ □

6.2.33 Let the angle $\widehat{xOy}$ be given and let A be a point in its interior. Construct a triangle ABC with $B \in Ox$, $C \in Oy$, $\widehat{BAC} = \widehat{\omega}$ such that $AB \cdot AC = k^2$, where k is the length of a given straight line segment and $\widehat{\omega}$ is a given angle.

Proof We will solve the problem in two steps. We will first provide an analysis and then we will proceed with the construction of the triangle ABC subject to the given conditions.

Analysis. Consider $AB \le AC$. Let us assume that the required triangle has been constructed. The fact that the product $AB \cdot AC$ is constant does really help us to prove that in due motion the angle $\widehat{A}$ remains constant (in measure). Therefore, if we consider a point $D \in AC$ such that the equality $AD = AB$ holds, the isosceles triangle ABD remains similar to itself, which means it preserves its angles. This property is useful for the determination of a certain motion of D. This motion is

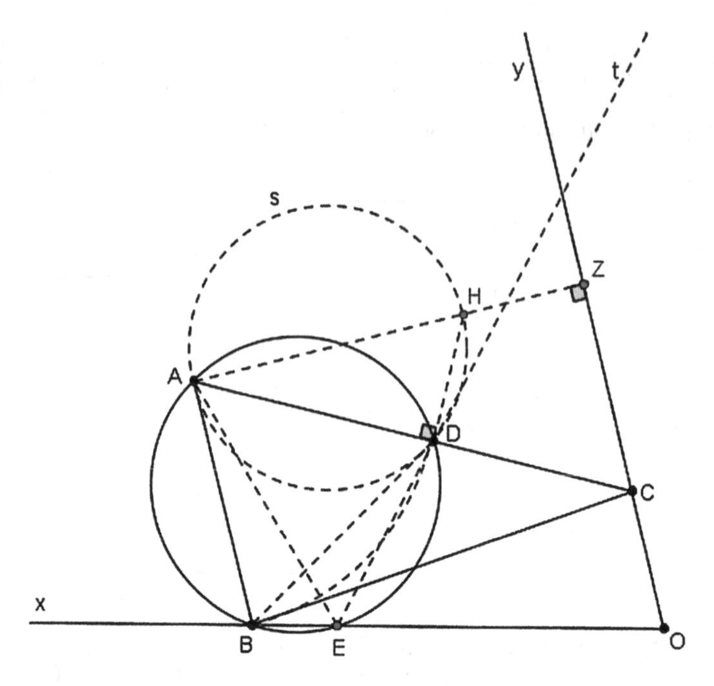

Fig. 6.53 Illustration of Problem 6.2.33

generated from the motion of the point C along the constant straight line Oy (see Fig. 6.53).

We consider the circumscribed circle of the triangle ABD. If this circle has a common point E with the straight semiline Ox then from the isosceles triangle ABD we deduce

$$\widehat{AEB} = \widehat{ADB} = 90^\circ - \frac{\widehat{\omega}}{2}. \tag{6.226}$$

Then, the point E is a constant point on Ox. But since

$$\widehat{DEO} = \widehat{\omega}, \tag{6.227}$$

it follows that the point D is moving on a constant straight semiline t, where t is passing through the point E and forms an angle $\widehat{\omega}$ with Ox. Let us denote this straight semiline by Et. Furthermore, by the assumption we made, we get

$$AD \cdot AC = AB \cdot AC = k^2,$$

and thus the point D has to belong, apart from the semiline Et, to the inverse figure of the Oy axis with the inversion with center A and power k^2. In this way, we have determined the point D, and consequently also the point C, as the intersection of two constant lines.

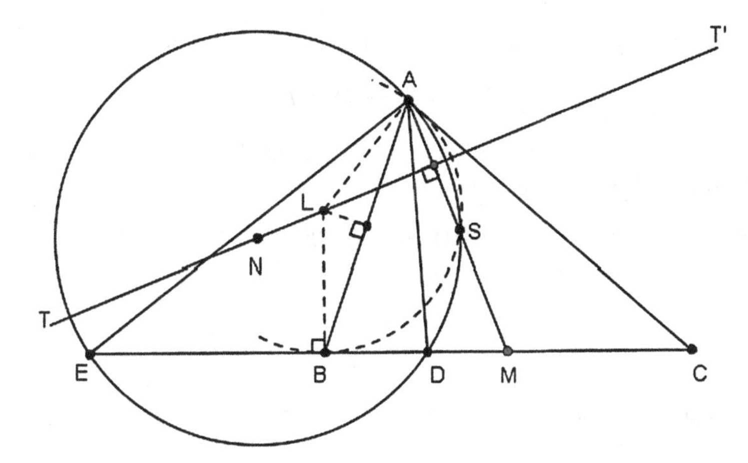

Fig. 6.54 Illustration of Problem 6.2.34

Construction. We define the point $E \in Ox$ such that

$$\widehat{AEx} = 90^\circ - \frac{\widehat{\omega}}{2} \tag{6.228}$$

and we form the angle

$$\widehat{xEt} = 180^\circ - \widehat{\omega}.$$

Let us draw AZ such that $AZ \perp Oy$ and on the straight semiline AZ we construct a point H such that

$$AZ \cdot AH = AD \cdot AC = AB \cdot AC = k^2. \tag{6.229}$$

Hence, the intersection of the circle of diameter AH with the straight semiline Et determines the point D. We can construct the point $B \in Ox$ such that

$$\widehat{BAD} = \widehat{\omega},$$

and the point C is determined as the intersection of the straight lines AD and Oy. The triangle we have thus constructed satisfies the given requirements. □

6.2.34 Let a triangle ABC with $BC = a$, $AC = b$, $AB = c$ and a point D in the interior of the side BC be given. Let E be the harmonic conjugate of D with respect to the points B and C. Determine the geometrical locus of the center of the circumferences DEA when D is moving along the side BC.

Analysis Let D be any point in side BC, different from the midpoint of the side BC and let E be its harmonic conjugate. Let N be the center of the circumference ADE. This is obviously a point belonging to the geometrical locus under investigation. Let S be the other point of intersection of the circumference ADE with the median AM of the triangle ABC (see Fig. 6.54).

Using the power of a point with respect to a circle, we obtain

$$MA \cdot MS = MD \cdot ME. \tag{6.230}$$

A necessary and sufficient condition for the points D, E to be harmonic conjugates of B, C, when M is the midpoint of BC, is

$$MD \cdot ME = MB^2. \tag{6.231}$$

However,

$$MB = \frac{a}{2},$$

where $a = BC$ and thus

$$MD \cdot ME = \frac{a^2}{4}. \tag{6.232}$$

Using (6.231) and (6.232), we obtain

$$MS \cdot MA = \frac{a^2}{4},$$

that is,

$$MS = \frac{a^2}{4MA}, \tag{6.233}$$

which is a constant.

Because of the fact that the point M is constant, it follows that the point S has constant position as well. Since the point A is given, it is evident that the point N should belong to the perpendicular bisector TT' of the straight line segment AS.

Construction of the geometrical locus We determine a point S in the median AM of the triangle ABC such that

$$MS \cdot MA = \left(\frac{a}{2}\right)^2. \tag{6.234}$$

We draw the perpendicular at the point B and we then determine the perpendicular bisector of the side AB. Let L be the point of intersection of the these two straight lines. With center at the point L and radius LA, where $LA = LB$, we draw a circle. The intersection of this circle with the median AM is the point S. Thus the point S is constructed. Therefore, the perpendicular bisector of the straight line segment AS is the geometrical locus of the point N.

Proof Let N be any point of the straight line that has been constructed. With center at the point N and radius NA we draw the circumference which passes through the point S and intersects the straight line BC at the points D, E. The point D is an internal point of the straight line segment BC and the point E is an external point of BC. The relation

$$MB^2 = MS \cdot MA = MD \cdot ME$$

holds true. This provides a necessary and sufficient condition for the points B, C, D, and E to be harmonic conjugates.

Remark 6.5 A necessary and sufficient condition for the four points B, C, D, E to form a harmonic quadruple is the following:

$$\frac{2}{BC} = \frac{1}{CE} + \frac{1}{CD}. \tag{6.235}$$

We can thus conclude that the harmonic conjugate of the midpoint is a point at infinity. It follows that the straight line TT' is the required geometrical locus. It is evident that the geometrical locus depends on the position of the point S. Therefore, we distinguish the following cases:

- If $AM > BC/2$, then S is in the interior of the triangle.
- If $AM < BC/2$, then S is in the exterior of the triangle.
- If $AM = BC/2$, then $S \equiv A$. □

6.3 Geometric Inequalities

6.3.1 Consider the triangle ABC and let H_1, H_2, H_3 be the intersection points of the altitudes AA_1, BB_1, CC_1, with the circumscribed circle of the triangle ABC, respectively. Show that

$$\frac{H_2H_3^2}{BC^2} + \frac{H_3H_1^2}{CA^2} + \frac{H_1H_2^2}{AB^2} \geq 3. \tag{6.236}$$

First solution We know that the symmetrical points of the orthocenter H of the triangle ABC with respect to the straight lines BC, CA, AB are the points H_1, H_2, H_3, respectively, which belong to the circumscribed circle of the triangle ABC (see Fig. 6.55). Thus

$$H_2H_3 = 2B_1C_1, \tag{6.237}$$

$$H_3H_1 = 2C_1A_1, \tag{6.238}$$

$$H_1H_2 = 2B_1A_1. \tag{6.239}$$

It is enough to show that

$$\frac{B_1C_1^2}{BC^2} + \frac{C_1A_1^2}{CA^2} + \frac{A_1B_1^2}{AB^2} \geq \frac{3}{4}. \tag{6.240}$$

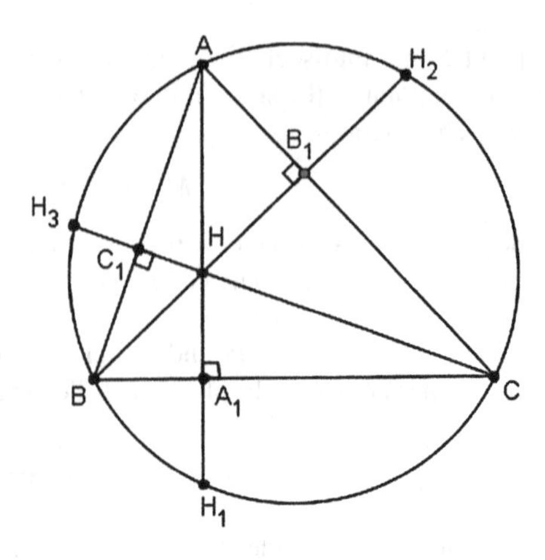

Fig. 6.55 Illustration of Problem 6.3.1

It follows that if $\widehat{A} \le 90^\circ$, then

$$AB_1 = \frac{AC^2 + AB^2 - BC^2}{2AC}, \tag{6.241}$$

and if $\widehat{A} > 90^\circ$, then

$$AB_1 = \frac{BC^2 - AC^2 - AB^2}{2AC}. \tag{6.242}$$

Therefore,

$$AB_1 = \pm\frac{AC^2 + AB^2 - BC^2}{2AC}. \tag{6.243}$$

Since the triangles AB_1C_1 and ABC are similar, we have

$$\frac{B_1C_1}{BC} = \frac{AB_1}{AB}, \tag{6.244}$$

and thus

$$\frac{B_1C_1^2}{BC^2} = \frac{AB_1^2}{AB^2} = \frac{(AC^2 + AB^2 - BC^2)^2}{4AC^2 \cdot AB^2}. \tag{6.245}$$

Similarly, we obtain

$$\frac{A_1C_1^2}{CA^2} = \frac{(AB^2 + BC^2 - AC^2)^2}{4AB^2 \cdot BC^2}, \tag{6.246}$$

and

$$\frac{A_1B_1^2}{AB^2} = \frac{(BC^2 + AC^2 - AB^2)^2}{4BC^2 \cdot AC^2}. \tag{6.247}$$

Therefore, it is enough to show that

$$\frac{BC^2 \cdot (AC^2 + AB^2 - BC^2)^2}{4AB^2 \cdot BC^2 \cdot CA^2} + \frac{AC^2 \cdot (AB^2 + BC^2 - AC^2)^2}{4AB^2 \cdot BC^2 \cdot CA^2} + \frac{AB^2 \cdot (BC^2 + AC^2 - AB^2)^2}{4AB^2 \cdot BC^2 \cdot CA^2} \geq \frac{3}{4}. \tag{6.248}$$

Without loss of generality, we consider $BC \geq AC \geq AB$. It can easily be seen (and is left as an exercise to the reader) that

$$\begin{aligned} &BC^2 \cdot \left(BC^2 - AC^2\right) \cdot \left(BC^2 - AB^2\right) \\ &\quad + AC^2 \cdot \left(AC^2 - AB^2\right) \cdot \left(AC^2 - BC^2\right) \\ &\quad + AB^2 \cdot \left(AB^2 - BC^2\right) \cdot \left(AB^2 - AC^2\right) \geq 0. \end{aligned} \tag{6.249}$$

We have

$$\begin{aligned} &BC^2 \cdot \left(BC^2 - AC^2\right) \cdot \left(BC^2 - AB^2\right) \\ &\quad + AC^2 \cdot \left(AC^2 - AB^2\right) \cdot \left(AC^2 - BC^2\right) \\ &\quad + AB^2 \cdot \left(AB^2 - BC^2\right) \cdot \left(AB^2 - AC^2\right) \\ &\geq 3BC^2 \cdot AC^2 \cdot AB^2, \end{aligned} \tag{6.250}$$

or

$$\begin{aligned} &\frac{BC^2 \cdot (AC^2 + AB^2 - BC^2)^2}{4AB^2 \cdot BC^2 \cdot CA^2} \\ &\quad + \frac{AC^2 \cdot (AB^2 + BC^2 - AC^2)^2}{4AB^2 \cdot BC^2 \cdot CA^2} \\ &\quad + \frac{AB^2 \cdot (BC^2 + AC^2 - AB^2)^2}{4AB^2 \cdot BC^2 \cdot CA^2} \geq \frac{3}{4}, \end{aligned} \tag{6.251}$$

which is actually (6.236).

Second solution (by Nicuşor Minculete and Cătălin Barbu) Since

$$\widehat{H_3AB} = 90^\circ - \widehat{B}, \qquad \widehat{H_2AC} = 90^\circ - \widehat{C},$$

we deduce that

$$\widehat{H_3AH_2} = 2\hat{A}.$$

Therefore, we have

$$H_2H_3 = 2R \sin \widehat{A} = 2a \cos \hat{A}.$$

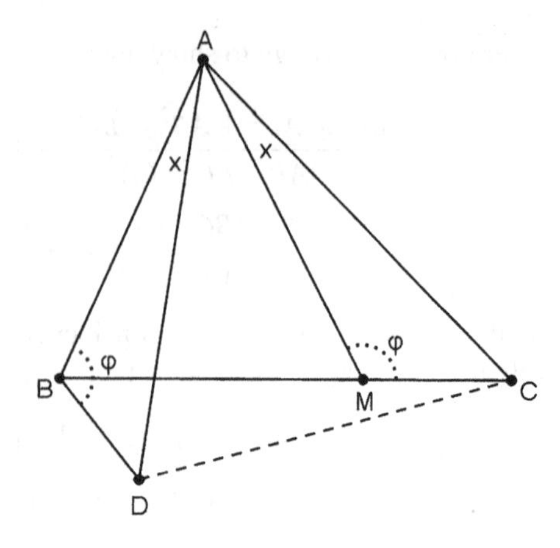

Fig. 6.56 Illustration of Problem 6.3.2

The inequality (6.248) becomes

$$\cos^2 \widehat{A} + \cos^2 \widehat{B} + \cos^2 \widehat{C} \geq \frac{3}{4}, \tag{6.252}$$

which is equivalent to

$$\sum_{\text{cycl}} \cos^2 \widehat{A} = 3 - \sum_{\text{cycl}} \sin^2 \widehat{A} = 3 - \frac{a^2+b^2+c^2}{4R^2} \geq \frac{3}{4}, \tag{6.253}$$

so

$$9R^2 \geq a^2 + b^2 + c^2,$$

which is true because

$$R^2 - \frac{a^2+b^2+c^2}{9} = OG^2 \geq 0,$$

where O is the center of the circumscribed circle of the triangle ABC and G is the centroid of the triangle ABC. □

6.3.2 Let ABC be a triangle with $AB = c$, $BC = a$ and $CA = b$, and let d_a, d_b, d_c be its internal angle bisectors. Show that

$$\frac{1}{d_a} + \frac{1}{d_b} + \frac{1}{d_c} > \frac{1}{a} + \frac{1}{b} + \frac{1}{c}. \tag{6.254}$$

First solution Let M be an interior point of BC and consider a point D in the plane of the triangle ABC such that (see Fig. 6.56)

$$\widehat{BAD} = \widehat{MAC} \quad \text{and} \quad \widehat{ABD} = \widehat{AMC}. \tag{6.255}$$

Because of the fact that

$$\widehat{AMC} > \widehat{B}, \tag{6.256}$$

the side BD lies outside of the triangle AMC. Since the triangles ABD and AMC are similar, we have

$$\frac{AB}{AM} = \frac{BD}{MC}. \tag{6.257}$$

Thus

$$AB \cdot MC = AM \cdot BD. \tag{6.258}$$

We also obtain

$$\frac{AB}{AM} = \frac{AD}{AC}. \tag{6.259}$$

Since

$$\widehat{BAM} = \widehat{DAC} \tag{6.260}$$

and because of (6.259), the triangles ABM and ADC are similar. Thus

$$\frac{AM}{AC} = \frac{MB}{DC},$$

which implies that

$$AC \cdot MB = DC \cdot AM. \tag{6.261}$$

Therefore,

$$AB \cdot MC + AC \cdot MB = AM(BD + DC), \tag{6.262}$$

and hence

$$AB \cdot MC + AC \cdot MB > AM \cdot BC. \tag{6.263}$$

If AM is the bisector d_a, then

$$BM = \frac{ac}{b+c} \tag{6.264}$$

and

$$MC = \frac{ab}{b+c}. \tag{6.265}$$

Thus

$$\frac{abc}{b+c} + \frac{abc}{b+c} > a \cdot d_a, \tag{6.266}$$

and therefore

$$d_a < \frac{2bc}{b+c}. \tag{6.267}$$

This implies that

$$\frac{1}{d_a} > \frac{b+c}{2bc}, \tag{6.268}$$

and so

$$\frac{1}{d_a} > \frac{1}{2}\left(\frac{1}{b} + \frac{1}{c}\right). \tag{6.269}$$

Similarly, we have

$$\frac{1}{d_b} > \frac{1}{2}\left(\frac{1}{a} + \frac{1}{c}\right) \tag{6.270}$$

and

$$\frac{1}{d_c} > \frac{1}{2}\left(\frac{1}{a} + \frac{1}{b}\right). \tag{6.271}$$

Adding inequalities (6.269), (6.270), and (6.271), we obtain

$$\frac{1}{d_a} + \frac{1}{d_b} + \frac{1}{d_c} > \frac{1}{2} \cdot 2\left(\frac{1}{a} + \frac{1}{b} + \frac{1}{c}\right), \tag{6.272}$$

which implies

$$\frac{1}{d_a} + \frac{1}{d_b} + \frac{1}{d_c} > \frac{1}{a} + \frac{1}{b} + \frac{1}{c}. \tag{6.273}$$

□

Second solution (by Nicuşor Minculete and Cătălin Barbu) Let AD be the internal angle bisector, where $D \in BC$. We apply Stewart's theorem and we obtain

$$AD^2 \cdot BC + BD \cdot DC \cdot BC = AB^2 \cdot DC + AC^2 \cdot BD.$$

It is easy to see that

$$AD = d_a, \qquad BD = \frac{ac}{b+c}, \qquad DC = \frac{ab}{b+c},$$

where $BC = a$, $AC = b$, and $AB = c$.

Therefore, we obtain

$$aAD^2 + \frac{a^3bc}{(b+c)^2} = \frac{abc^2}{b+c} + \frac{ab^2c}{b+c} = abc,$$

which implies the equality

$$\begin{aligned} AD^2 &= \frac{bc}{(b+c)^2}\left[(b+c)^2 - a^2\right] \\ &= \frac{bc}{(b+c)^2}(2bc\cos A + 2bc) \\ &= \frac{2b^2c^2}{(b+c)^2}(\cos A + 1) \\ &= \frac{4b^2c^2}{(b+c)^2}\cos^2\frac{\hat{A}}{2}. \end{aligned}$$

It follows that

$$d_a = \frac{2bc}{b+c}\cos\frac{A}{2} < \frac{2bc}{b+c}, \tag{6.274}$$

which implies the inequality

$$\frac{1}{d_a} > \frac{1}{2}\left(\frac{1}{b} + \frac{1}{c}\right).$$

In the analogous way, we deduce the inequalities

$$\frac{1}{d_b} > \frac{1}{2}\left(\frac{1}{a} + \frac{1}{c}\right)$$

and

$$\frac{1}{d_c} > \frac{1}{2}\left(\frac{1}{b} + \frac{1}{a}\right).$$

Combining we obtain the statement

$$\frac{1}{d_a} + \frac{1}{d_b} + \frac{1}{d_c} > \frac{1}{a} + \frac{1}{b} + \frac{1}{c}.$$

□

6.3.3 Let ABC be a triangle with $\widehat{C} > 10^\circ$ and $\widehat{B} = \widehat{C} + 10^\circ$. Consider a point E on AB such that $\widehat{ACE} = 10^\circ$ and let D be a point on AC such that $\widehat{DBA} = 15^\circ$. Let $Z \neq A$ be a point of intersection of the circumscribed circles of the triangles ABD and AEC. Show that

$$\widehat{ZBA} > \widehat{ZCA}.$$

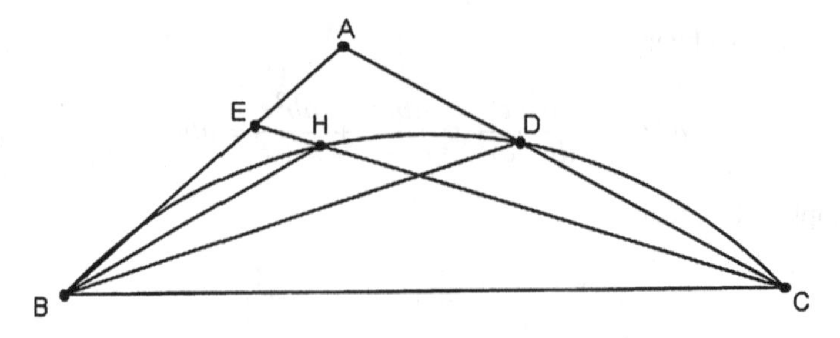

Fig. 6.57 Illustration of Problem 6.3.3

Solution We initially show that $BD < CE$ (see Fig. 6.57).

Since

$$\widehat{B} > \widehat{C}, \tag{6.275}$$

we have

$$AC > AB. \tag{6.276}$$

Also,

$$\widehat{BDC} = \widehat{A} + \widehat{DBA} = \widehat{A} + 15^\circ \tag{6.277}$$

and

$$\widehat{CEB} = \widehat{A} + 10^\circ. \tag{6.278}$$

From (6.277) and (6.278), we conclude that

$$\widehat{BDC} > \widehat{CEB}. \tag{6.279}$$

Therefore, the circle circumscribed to the triangle DBC intersects EC at a point H between E and C and thus

$$EC > HC. \tag{6.280}$$

We have

$$\begin{aligned}\widehat{HBC} &= \widehat{HBD} + \widehat{DBC} = \widehat{HCD} + \widehat{DBC} \\ &= 10^\circ + \widehat{B} - \widehat{DBA} = 10^\circ + \widehat{C} + 10^\circ - 15^\circ \qquad (6.281) \\ &= \widehat{C} + 5^\circ > \widehat{C}. \qquad (6.282)\end{aligned}$$

Therefore,

$$\widehat{HBC} > \widehat{C}, \tag{6.283}$$

and thus

$$HC > BD. \tag{6.284}$$

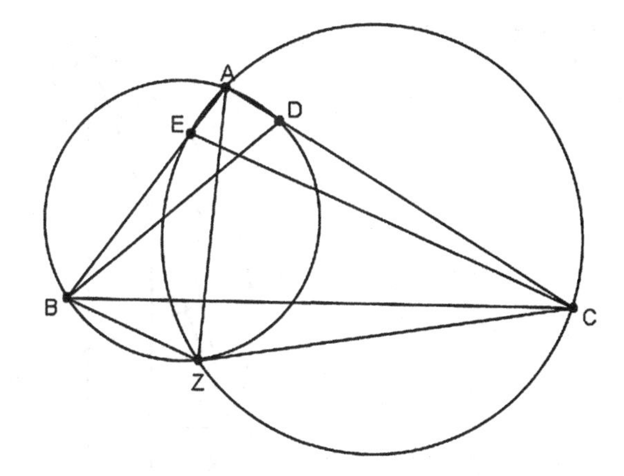

Fig. 6.58 Illustration of Problem 6.3.3

From Eqs. (6.282) and (6.284), it follows that

$$BD < EC. \tag{6.285}$$

The arc that subtends an angle of $180° - \widehat{A}$ corresponding to the chord DB is a set of points lying on a different half-plane than A with respect to BD. Thus, the arc lies inside the angle $\widehat{A}$. Also, the arc that subtends an angle of $180° - \widehat{A}$ corresponding to the chord CE is a set of points that lie on a different half-plane than A with respect to CE. So, it also lies inside the angle $\widehat{A}$. This means that the points B and C lie on opposite sides of the line containing the common chord AZ (see Fig. 6.58) and furthermore the point B lies outside the disk C_2, whereas the vertex C lies inside the disk C_1. Thus, from $\widehat{BAD} = \widehat{EAC}$ and $BD < EC$, we see that the radius of C_1 is smaller than the radius of C_2. Therefore,

$$\widehat{ZBA} > \widehat{ZCA}. \tag{6.286}$$

Comment If Z belongs to BC then

$$\widehat{ZBA} = \widehat{B} \quad \text{and} \quad \widehat{ZCA} = \widehat{C}, \tag{6.287}$$

with $\widehat{B} > \widehat{C}$ by hypothesis. □

6.3.4 Let ABC be a triangle of area S and D, E, F be points on the lines BC, CA, and AB, respectively. Suppose that the perpendicular lines at the points D, E, F to the lines BC, CA, and AB, respectively, intersect the circumcircle of ABC at the pairs of points (D_1, D_2), (E_1, E_2), and (F_1, F_2), respectively. Prove that

$$|D_1B \cdot D_1C - D_2B \cdot D_2C| \\ + |E_1C \cdot E_1A - E_2C \cdot E_2A| + |F_1A \cdot F_1B - F_2A \cdot F_2B| > 4S.$$

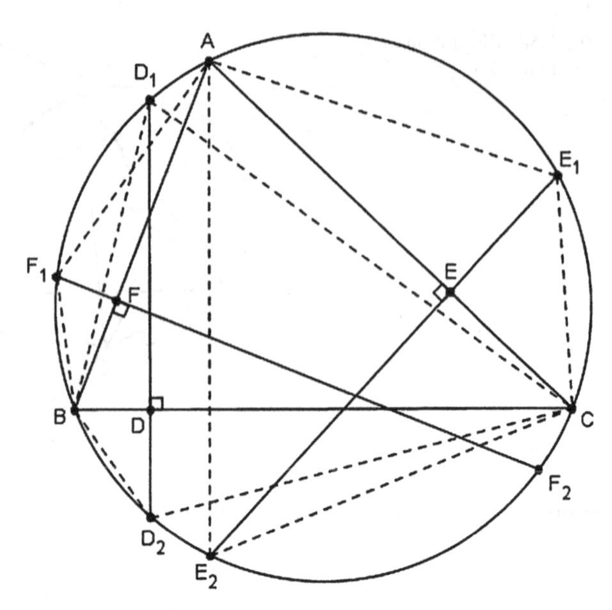

Fig. 6.59 Illustration of Problem 6.3.4

Solution We start with the following (see Fig. 6.59)

Lemma 6.10 *Suppose AB and D_1D_2 are perpendicular chords in a circle of center O. Then*

$$|S_{D_1AB} - S_{ABD_2}| = 2S_{AOB}. \tag{6.288}$$

Proof Let D_1' be the reflection of D_1 across AB. Then

$$\widehat{BAD_1'} = \widehat{BAD_1} \tag{6.289}$$

$$= \widehat{D_1D_2B}$$

$$= 90^\circ - \widehat{ABD_2}. \tag{6.290}$$

Hence

$$AD_1' \perp BD_2. \tag{6.291}$$

If BB' is the diameter of the circle, we infer that

$$B'D_2 \parallel AD_1' \quad \text{and} \quad AB' \parallel D_1D_2. \tag{6.292}$$

Thus the quadrilateral $AB'D_2D_1'$ is a parallelogram and

$$D_1'D_2 = AB' = 2OO', \tag{6.293}$$

where O' is the projection of O on AB. Consequently,

$$S_{ABD_2} - S_{ABD_1} = \frac{AB \cdot D_1' D_2}{2} = 2 \cdot S_{AOB}, \tag{6.294}$$

as desired. □

Now, apply the lemma successively for the pairs of perpendicular chords $BC \perp D_1D_2$, $CA \perp E_1E_2$, and $AB \perp F_1F_2$. It follows that

$$\begin{aligned} &|D_1B \cdot D_1C - D_2B \cdot D_2C| \\ &\geq |D_1B \cdot D_1C - D_2B \cdot D_2C| \cdot |\sin \widehat{BAC}| \\ &= |D_1B \cdot D_1C \cdot \sin \widehat{BAC} - D_2B \cdot D_2C \cdot \sin \widehat{BAC}| \\ &= 2 \cdot |S_{BCD_1} - S_{BCD_2}|, \end{aligned}$$

since

$$\widehat{BAC} = \widehat{BD_1C} = 180^\circ - \widehat{BD_2C},$$

which implies that

$$\sin \widehat{BAC} = \sin \widehat{BD_1C} = \sin \widehat{BD_2C}.$$

Therefore,

$$|D_1B \cdot D_1C - D_2B \cdot D_2C| \geq 4S_{BOC}. \tag{6.295}$$

Similarly,

$$|E_1C \cdot E_1A - E_2C \cdot E_2A| \geq 4S_{AOC} \tag{6.296}$$

and

$$|F_1A \cdot F_1B - F_2A \cdot F_2B| \geq 4S_{AOB}. \tag{6.297}$$

Adding inequalities (6.295), (6.296) and (6.297) implies the desired result since the equality holds only if

$$\sin \widehat{BAC} = \sin \widehat{CBA} = \sin \widehat{ACB} = 1, \tag{6.298}$$

which is impossible. □

6.3.5 Let ABC be an equilateral triangle and D, E be points on its sides AB and AC, respectively. Let F, G be points on the segments AE and AD, respectively, such that the lines DF and EG bisect the angles $\widehat{EDA}$ and $\widehat{AED}$, respectively. Prove that

$$S_{DEF} + S_{DEG} \leq S_{ABC}. \tag{6.299}$$

When does the equality hold?

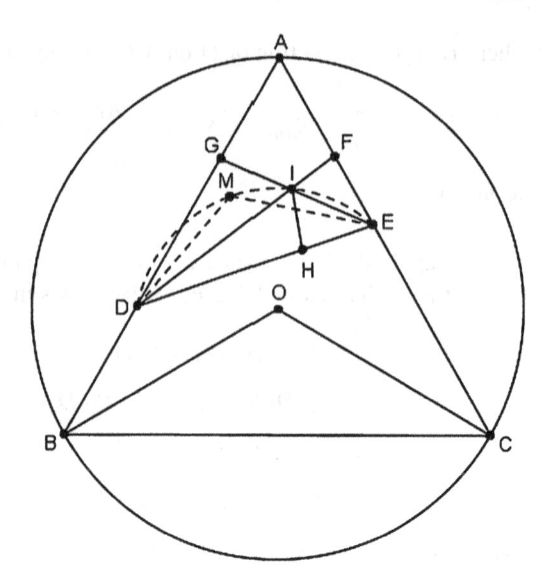

Fig. 6.60 Illustration of Problem 6.3.5

Solution We have

$$\widehat{BAC} = 60^\circ. \tag{6.300}$$

This implies that the angle $\widehat{DIE}$ is known, where I is the point of intersection of the bisectors DF and EG. We obtain (see Fig. 6.60)

$$\widehat{DIE} = 120^\circ,$$

since

$$\begin{aligned}\widehat{DIE} &= 180^\circ - \frac{\widehat{EDA}}{2} - \frac{\widehat{AED}}{2}\\ &= 180^\circ - \frac{\widehat{EDA} + \widehat{AED}}{2}\\ &= 180^\circ - \frac{120^\circ}{2} = 120^\circ.\end{aligned}$$

Hence,

$$\widehat{GID} = \widehat{EIF} = 60^\circ,$$

which implies that when IH bisects the angle $\widehat{DIE} = 120^\circ$, and we have

$$GDI = IDH \quad \text{and} \quad IEF = IEH.$$

Therefore,

$$S_{DEF} + S_{DEG} = 3S_{IDE}. \tag{6.301}$$

We shall show that

$$3S_{IDE} \leq S_{ABC}. \tag{6.302}$$

If DE moves in such a way that its length remains constant, then the position of DE for which we obtain the maximum area S_{IDE}, occurs when the segment DE is parallel to the base BC.

This is the case because the motion, just described, creates the triangles IDE when DE has constant position and constant length and the points I move on the arc whose points are the vertices of $120°$ angles subtending the chord DE. The position that gives the maximum area S_{IDE} of the triangle IDE is when I takes the place of the midpoint of this arc, and therefore when the triangle ADE becomes equilateral. If O is the circumcenter of the triangle ABC, then the triangles IDE and OBC are similar and

$$DE \leq BC. \tag{6.303}$$

Hence

$$S_{IDE} \leq S_{OBC},$$

and therefore

$$3S_{IDE} \leq S_{OBC}.$$

Thus

$$S_{ADE} \leq S_{ABC},$$

where equality holds in the case when the point D coincides with the point B and the point E coincides with the point C. □

6.3.6 Let PQR be a triangle. Prove that

$$\frac{1}{y+z-x}+\frac{1}{z+x-y}+\frac{1}{x+y-z} \geq \frac{1}{x}+\frac{1}{y}+\frac{1}{z}, \tag{6.304}$$

where

$$x=\sqrt{\sqrt[3]{QR^2}+\sqrt[5]{QR^2}}, \qquad y=\sqrt{\sqrt[3]{PR^2}+\sqrt[5]{PR^2}} \text{ and } z=\sqrt{\sqrt[3]{PQ^2}+\sqrt[5]{PQ^2}}.$$

Solution The proof is based on the following two lemmas:

Lemma 6.11 *Let $a=BC$, $b=AC$, $c=AB$ be the lengths of the sides of a triangle ABC. Then $\sqrt[n]{a}$, $\sqrt[n]{b}$, $\sqrt[n]{c}$ are also lengths of the sides of a triangle.*

Indeed, since a, b, c are the lengths of the sides of a triangle, it holds:

$$a+b>c, \qquad a+c>b, \quad \text{and} \quad b+c>a.$$

However,

$$\sqrt[n]{a}+\sqrt[n]{b}>\sqrt[n]{c}\Leftrightarrow a+b+M>c,$$

where

$$M=\left(\sqrt[n]{a}+\sqrt[n]{b}\right)^n-(a+b)$$

and similarly for b, c, a and c, a, b.

Lemma 6.12 *Let a, b, c and k, l, m be lengths of the sides of certain triangles. Then,*

$$\sqrt{a^2+k^2},\quad \sqrt{b^2+l^2},\quad \sqrt{c^2+m^2}$$

are also lengths of the sides of a triangle.

Proof Assume that

$$\sqrt{a^2+k^2},\quad \sqrt{b^2+l^2},\quad \sqrt{c^2+m^2}$$

are the lengths of the sides of a triangle. Then

$$\sqrt{a^2+k^2}+\sqrt{b^2+l^2}>\sqrt{c^2+m^2}$$

if and only if

$$a^2+k^2+b^2+l^2+2\sqrt{\left(a^2+k^2\right)\left(b^2+l^2\right)}>c^2+m^2.$$

Indeed, using the Cauchy–Schwarz–Buniakowski inequality, we get

$$\begin{aligned}a^2+k^2+b^2+l^2+2\sqrt{\left(a^2+k^2\right)\left(b^2+l^2\right)}&\geq a^2+k^2+b^2+l^2+2(ab+kl)\\&=(a+b)^2+(k+l)^2\\&>c^2+m^2.\end{aligned}$$

Similarly, we derive the other two inequalities. □

Since PQ, QR, and RP are the sides of a triangle, it follows, using the above two lemmas, that x, y, and z are lengths of the sides of a triangle. Hence, there exist positive real numbers k_1, m_1, n_1 such that

$$x=k_1+m_1,\qquad y=m_1+n_1,\qquad z=k_1+n_1$$

and inequality (6.304) assumes the form

$$\frac{1}{k_1}+\frac{1}{m_1}+\frac{1}{n_1}\geq 2\left(\frac{1}{k_1+m_1}+\frac{1}{m_1+n_1}+\frac{1}{n_1+k_1}\right), \tag{6.305}$$

which is easily verified by applying the inequalities

$$\frac{1}{k_1} + \frac{1}{m_1} \geq \frac{4}{k_1 + m_1},$$
$$\frac{1}{m_1} + \frac{1}{n_1} \geq \frac{4}{m_1 + n_1},$$
$$\frac{1}{n_1} + \frac{1}{k_1} \geq \frac{4}{k_1 + n_1}.$$

The equality holds true for the case of an equilateral triangle. □

6.3.7 The point O is considered inside the convex quadrilateral $ABCD$ of area S. Suppose that K, L, M, N are interior points (see Fig. 6.61) of the sides AB, BC, CD, and DA, respectively. If $OKBL$ and $OMDN$ are parallelograms of areas S_1 and S_2, respectively, prove that

$$\sqrt{S_1} + \sqrt{S_2} < 1.25\sqrt{S}, \tag{6.306}$$

$$\sqrt{S_1} + \sqrt{S_2} < C_0\sqrt{S}, \tag{6.307}$$

where

$$C_0 = \max_{0<\alpha<\pi/4} \frac{\sin(2\alpha + \frac{\pi}{4})}{\cos\alpha}.$$

(Proposed by Nairi Sedrakyan [88], Armenia)

Solution We can assume, without loss of generality, that the points O and D are not on different sides of the line AC. Assume

$$S_{ABC} = a, \qquad S_{ACD} = b, \qquad S_{OAC} = x,$$

$$S_{OKB} = S_{OBL} = S_{KLB} = \frac{S_1}{2},$$

and

$$\frac{S_{OKB}}{S_{OAB}} \cdot \frac{S_{OBL}}{S_{OBC}} = \frac{KB}{AB} \cdot \frac{BL}{BC} = \frac{S_{KBL}}{S_{ABC}}. \tag{6.308}$$

Then

$$S_1 = \frac{2S_{OAB} \cdot S_{OBC}}{a}.$$

We also get

$$S_2 = \frac{2S_{OAD} \cdot S_{OCD}}{b}.$$

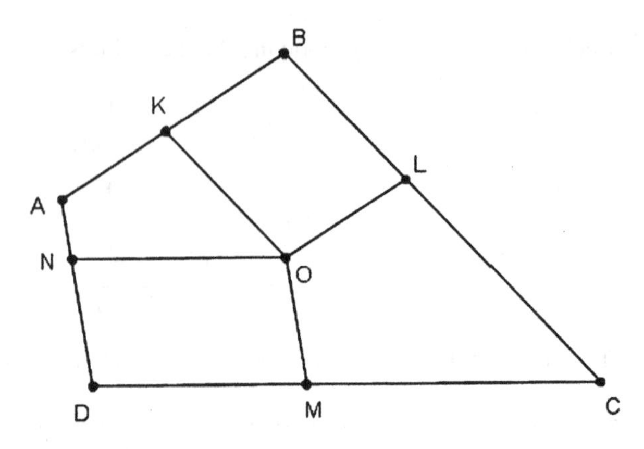

Fig. 6.61 Illustration of Problem 6.3.7

Hence

$$\sqrt{S_1}+\sqrt{S_2} \leq \frac{S_{OAB}+S_{OBC}}{\sqrt{2a}}+\frac{S_{OAD}+S_{OCD}}{\sqrt{2b}}=\frac{a+x}{\sqrt{2a}}+\frac{b-x}{\sqrt{2b}}$$
$$=\frac{\sqrt{a}+\sqrt{b}}{\sqrt{b}}-\frac{\sqrt{a}-\sqrt{b}}{\sqrt{2ab}}x. \tag{6.309}$$

For $a \geq b$, we have

$$\sqrt{S_1}+\sqrt{S_2} \leq \frac{\sqrt{a}+\sqrt{b}}{\sqrt{2}} \leq \sqrt{a+b}=\sqrt{S}.$$

For $a < b$, it follows that the point O cannot be outside the parallelogram $ABCE$, and thus $x \leq a$. Therefore,

$$\sqrt{S_1}+\sqrt{S_2} \leq \frac{\sqrt{a}+\sqrt{b}}{\sqrt{2}}-\frac{\sqrt{a}-\sqrt{b}}{\sqrt{2ab}}a=\frac{b+\sqrt{2ab}-a}{\sqrt{2b}}. \tag{6.310}$$

Let

$$\frac{a}{b}=\tan^2\alpha, \quad \alpha \in \left[0, \frac{\pi}{4}\right].$$

Then

$$\frac{b-\sqrt{2ab}-a}{\sqrt{2b}}/\sqrt{a+b}=\frac{\sin(2\alpha+\frac{\pi}{4})}{\cos\alpha} \leq C_0.$$

Consequently,

$$\sqrt{S_1}+\sqrt{S_2} \leq \frac{b+\sqrt{2ab}-a}{\sqrt{2b}} \leq C_0\sqrt{S}$$

when

$$\alpha \in \left[\frac{\pi}{4}, \frac{\sin(2\alpha+\frac{\pi}{4})}{\cos\alpha}-1\right],$$

that is, $C_0 \geq 1$. Thus, in all cases

$$\sqrt{S_1} + \sqrt{S_2} \leq C_0\sqrt{S}.$$

If for the quadrilateral the following condition holds true

$$AB = BC \cdot AD = CD \cdot \frac{S_{ABC}}{\tan\alpha_0},$$

where

$$C_0 = \frac{\sin(2\alpha_0 + \frac{\pi}{4})}{\cos\alpha_0},$$

and $ABCO$ is a parallelogram, then

$$\sqrt{S_1} + \sqrt{S_2} = C_0\sqrt{S}.$$

This proves the assertion (6.307).

To prove inequality (6.306), it is sufficient to verify the property that if $0 \leq \alpha \leq \frac{\pi}{4}$ then

$$\sin\left(2\alpha + \frac{\pi}{4}\right) < 1.25\cos\alpha.$$

Indeed, let $\phi \in [0, \frac{\pi}{4}]$ and $\cos\phi = \frac{4}{5}$, then, if $0 \leq \alpha < \phi$, it follows that

$$\sin(2\alpha + \phi) \leq 1 = \frac{5}{4}\cos\phi.$$

Furthermore, if $\phi \leq \alpha \leq \frac{\pi}{4}$, then

$$\tan\phi = \frac{3}{4} > \sqrt{2} - 1 = \tan\frac{\pi}{8},$$

hence

$$\phi > \frac{\pi}{8}$$

and

$$\sin\left(2\alpha + \frac{\pi}{4}\right) \leq \sin\left(2\phi + \frac{\pi}{4}\right) = \frac{\sqrt{2}}{2} \cdot \frac{31}{25} < \frac{\sqrt{2}}{2} \cdot \frac{5}{4} \leq 1.25\cos\alpha. \tag{6.311}$$

□

Remark It can be proved that

$$\tan\alpha_0 = \sqrt[3]{\sqrt{2}+1} - \sqrt[3]{\sqrt{2}-1} = 0.59\ldots, \quad \text{while } C_0 = 1.11\ldots. \tag{6.312}$$

6.3.8 Let $ABCD$ be a quadrilateral with $\widehat{A} \geq 60°$. Prove that

$$AC^2 \leq 2\big(BC^2 + CD^2\big), \tag{6.313}$$

with equality, when $AB = AC$, $BC = CD$, and $\widehat{A} = 60°$.
(*Proposed by Titu Andreescu [6], USA*)

Solution (by Daniel Lasaosa, Spain) From Ptolemy's inequality, we have

$$AC \cdot BD \leq AB \cdot CD + BC \cdot DA.$$

The equality is attained if and only if the quadrilateral $ABCD$ is cyclic. Because of the fact that $\widehat{A} > 60°$ it follows that $\cos A < \frac{1}{2}$. However, by the cosine law, we get

$$BD^2 > AB^2 + AD^2 - AB \cdot AD.$$

Therefore,

$$AC < \frac{AB \cdot CD + BC \cdot DA}{\sqrt{AB^2 + AD^2 - AB \cdot AD}}. \tag{6.314}$$

It is enough to show that

$$\frac{(AB \cdot CD + BC \cdot DA)^2}{AB^2 + AD^2 - AB \cdot AD} \leq 2\big(BC^2 + CD^2\big). \tag{6.315}$$

The inequality (6.315) can be expressed as follows

$$\big(BC^2 + CD^2\big)(AB - AD)^2 + (AB \cdot BC - CD \cdot DA)^2 \geq 0. \tag{6.316}$$

This becomes an equality if and only if $AB = AD$ and $BC = CD$. This completes the proof. □

6.3.9 Let R and r be the circumradius and the inradius of the triangle ABC with sides of lengths a, b, c (see Fig. 6.62). Prove that

$$2 - 2\sum_{\text{cycl}} \left(\frac{a}{b+c}\right)^2 \leq \frac{r}{R}. \tag{6.317}$$

(*Proposed by Dorin Andrica [18], Romania*)

Solution (by Arkady Alt, California, USA) It is clear that

$$2 - 2\sum_{\text{cycl}} \left(\frac{a}{b+c}\right)^2 \leq \frac{r}{R}$$

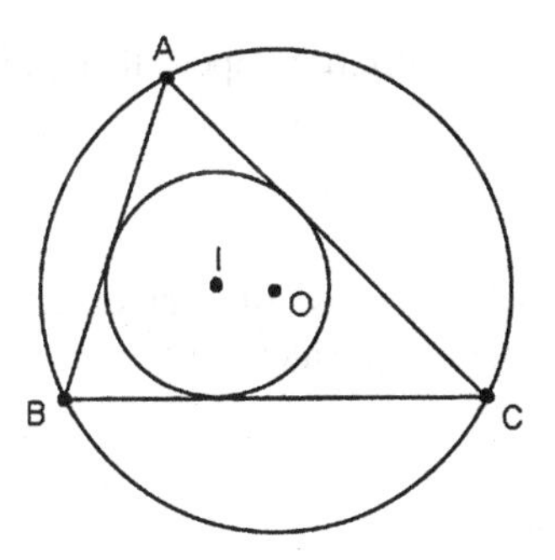

Fig. 6.62 Illustration of Problem 6.3.9

is equivalent to the following inequality:

$$6-2\sum_{\text{cycl}}\left(\frac{a}{b+c}\right)^2 \le 4+\frac{r}{R}.$$

Therefore,

$$2\left(3-\sum_{\text{cycl}}\left(\frac{a}{b+c}\right)^2\right) \le 4+\frac{r}{R},$$

and hence

$$2\sum_{\text{cycl}}\frac{(b+c)^2-a^2}{(b+c)^2} \le 4+\frac{r}{R}. \quad (6.318)$$

Because of the fact that

$$\cos A+\cos B+\cos C = 1+\frac{r}{R}$$

as well as

$$\frac{1}{(b+c)^2} \le \frac{1}{4bc},$$

it follows that

$$\frac{(b+c)^2-a^2}{2bc} = 1+\cos\widehat{A}.$$

Thus we obtain

$$\begin{aligned} 2\sum_{\text{cycl}}\frac{(b+c)^2-a^2}{(b+c)^2} &\le \sum_{\text{cycl}}\frac{(b+c)^2-a^2}{2bc} \\ &= \sum_{\text{cycl}}(1+\cos A) \\ &= 4+\frac{r}{R}. \end{aligned} \quad (6.319)$$

□

Remark 6.6 Suppose that l_a, l_b, l_c are the angle bisectors of a triangle ABC. Since

$$\frac{(b+c)^2-a^2}{(b+c)^2}=\frac{al_a^2}{abc}$$

(the proof is left as an exercise to the reader), the inequality (6.319) can be written in the equivalent form

$$2\sum_{\text{cycl}}\frac{al_a^2}{abc}\le 4+\frac{r}{R},$$

or

$$2\sum_{\text{cycl}}\frac{al_a^2}{4Rrs}\le 4+\frac{r}{R},$$

or

$$\frac{al_a^2+bl_b^2+cl_c^2}{a+b+c}\le r(4R+r). \tag{6.320}$$

Second proof (by Nicuşor Minculete) In the book, N. Minculete, Geometric Equalities and Inequalities in the triangle, Editura Eurocarpatica, Sfârtu Gheorghe, 2003 (in Romanian), the following inequality is proved

$$\frac{a}{b+c}\ge \sin\frac{A}{2}\ge\sqrt{\frac{2r}{R}}\cdot\frac{a}{b+c}. \tag{6.321}$$

Thus, we deduce

$$\sum_{\text{cycl}}\left(\frac{a}{b+c}\right)^2\ge\sum_{\text{cycl}}\sin^2\frac{A}{2}\ge\frac{2r}{R}\sum_{\text{cycl}}\left(\frac{a}{b+c}\right)^2. \tag{6.322}$$

But

$$\begin{aligned}\sum_{\text{cycl}}\sin^2\frac{A}{2}&=\frac{1}{2}\big(3-(\cos A+\cos B+\cos C)\big)\\&=1-\frac{r}{2R}.\end{aligned}$$

Therefore, we have

$$\begin{aligned}\sum_{\text{cycl}}\left(\frac{a}{b+c}\right)^2&\ge 1-\frac{r}{2R}\\&\ge\frac{2r}{R}\sum_{\text{cycl}}\left(\frac{a}{b+c}\right)^2.\end{aligned}$$

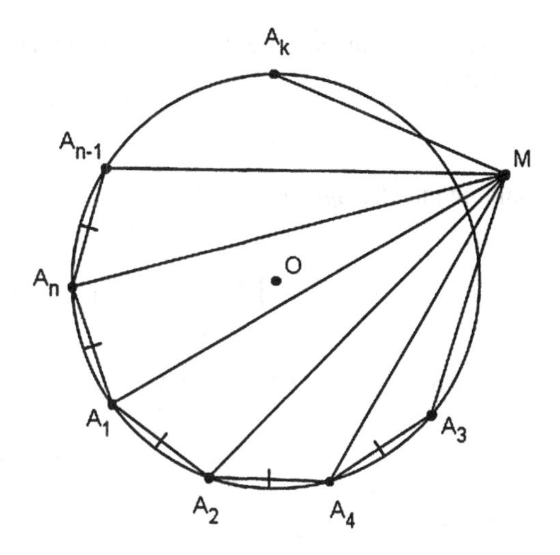

Fig. 6.63 Illustration of Problem 6.3.10

It follows that

$$\frac{5}{2}-\frac{R}{r}\leq 2-2\sum_{\text{cycl}}\left(\frac{a}{b+c}\right)^2\leq\frac{r}{R}.$$

□

6.3.10 Let $A_1A_2\dots A_n$ be a regular n-gon inscribed in a circle of center O and radius R. Prove that for each point M in the plane of the n-gon the following inequality holds

$$\prod_{k=1}^{n} MA_k\leq\left(OM^2+R^2\right)^{n/2}. \tag{6.323}$$

(Proposed by Dorin Andrica [15], Romania)

Solution (by Samin Riasat, Bangladesh) Let O be the origin in the complex plane. Without loss of generality, let us assume that $R=1$. Assume that (see Fig. 6.63)

$$\omega=\exp\left(\frac{2\pi i}{n}\right)$$

is the nth root of unity, and let the complex numbers $\omega,\omega^2,\dots,\omega^n,x$ correspond to the points $A_1,A_2,\dots,A_n$, and M, respectively, in the complex plane.

It follows that the inequality (6.323) is equivalent to the inequality

$$\prod_{k=1}^{n}\left|x-\omega^k\right|\leq\sqrt{\left(|x|^2+1\right)^n}. \tag{6.324}$$

Because of the fact that the complex numbers $\omega, \omega^2, \ldots, \omega^n$ are the roots of the equation

$$z^n - 1 = 0,$$

applying the triangle inequality, we get

$$\prod_{k=1}^{n} |x - \omega^k| = |x^n - 1| \leq |x|^n + 1. \tag{6.325}$$

Therefore, it suffices to prove that

$$\left(|x|^n + 1\right)^2 \leq \left(|x|^2 + 1\right)^2,$$

that is,

$$2|x|^n \leq \sum_{k=1}^{n-1} \left(\frac{n!}{k!(n-k)!}\right)|x|^{2k}. \tag{6.326}$$

This is a consequence of the arithmetic mean—geometric mean inequality (Cauchy's inequality) since

$$\begin{aligned}\sum_{k=1}^{n-1} \frac{n!}{k!(n-k)!}|x|^{2k} &\geq n|x|^2 + n|x|^{2n-2}\\ &\geq 2n|x|^n \geq 2|x|^n\end{aligned} \tag{6.327}$$

and $n \geq 3$. This completes the proof of the claim. □

The equality holds if and only if $|x| = 0$, that is, when $M \equiv O$.

Remark 6.7 The reader will find the book of T. Andreescu and D. Andrica [21] a very useful source for theory and problem-solving using complex numbers.

6.3.11 Let (K_1, a), (K_2, b), (K_3, c), (K_4, d) be four cyclic disks of a plane Π, having at least one common point. Let I be a point of their intersection. Let also O be a point in the plane Π such that

$$\min\{(OA), (OA'), (OB), (OB'), (OC), (OC'), (OD), (OD')\} \geq (OI) + 2\sqrt{2}, \tag{6.328}$$

where AA', BB', CC', DD' are the diameters of (K_1, a), (K_2, b), (K_3, c), and (K_4, d), respectively. Prove that

$$144 \cdot \left(a^4+b^4+c^4+d^4\right) \cdot \left(a^8+b^8+c^8+d^8\right)$$
$$\geq \left[\left(\frac{ab+cd}{2}\right)^2+\left(\frac{ad+bc}{2}\right)^2+\left(\frac{ac+bd}{2}\right)^2\right]$$
$$\cdot \left[(a+b)\cdot(c+d)+(a+d)\cdot(b+c)+(a+c)\cdot(b+d)\right]. \qquad (6.329)$$

Under what conditions does the equality in (6.329) hold?

Solution The equality is valid when the diameters of the circles are sides of a square with length equal to 1 and the intersection point I of its diameters coincides with O.

Let us consider

$$a_1 = 2a, \qquad b_1 = 2b, \qquad c_1 = 2c, \qquad d_1 = 2d.$$

Hence, in order to prove (6.329), it suffices to verify that

$$9 \cdot \left(a_1^4+b_1^4+c_1^4+d_1^4\right) \cdot \left(a_1^8+b_1^8+c_1^8+d_1^8\right)$$
$$\geq \left[(a_1b_1+c_1d_1)^2+(a_1d_1+b_1c_1)^2+(a_1c_1+b_1d_1)^2\right]$$
$$\cdot \left[(a_1+b_1)\cdot(c_1+d_1)+(a_1+d_1)\cdot(b_1+c_1)+(a_1+c_1)\cdot(b_1+d_1)\right]. \qquad (6.330)$$

However,

$$(DI) \geq \left|(OD)-(OI)\right| \geq \left|(OI)+\frac{\sqrt{2}}{2}-(OI)\right| = \frac{\sqrt{2}}{2}. \qquad (6.331)$$

Therefore,

$$(DI) \geq \frac{\sqrt{2}}{2}, \qquad (D'I) \geq \frac{\sqrt{2}}{2},$$
$$(AI) \geq \frac{\sqrt{2}}{2}, \qquad (A'I) \geq \frac{\sqrt{2}}{2},$$
$$(BI) \geq \frac{\sqrt{2}}{2}, \qquad (B'I) \geq \frac{\sqrt{2}}{2},$$
$$(CI) \geq \frac{\sqrt{2}}{2}, \qquad (C'I) \geq \frac{\sqrt{2}}{2}. \qquad (6.332)$$

The triangle IAA' satisfies the property $\widehat{AIA'} \geq 90^\circ$ since the point I is either in the interior of the cyclic disk or it belongs to the circumference with diameter AA'. Therefore,

$$\left(AA'\right)^2 \geq (IA)^2+\left(IA'\right)^2, \qquad (6.333)$$

where $(AA') = a_1$. Thus, $a_1^2 \geq 1$.

Similarly, from the inequalities

$$b_1^2 \geq 1, \qquad c_1^2 \geq 1, \qquad d_1^2 \geq 1, \tag{6.334}$$

it follows that

$$b_1 \geq 1, \qquad c_1 \geq 1, \qquad d_1 \geq 1. \tag{6.335}$$

Thus, we obtain

$$a_1^4 \geq a_1^2, \qquad b_1^4 \geq b_1^2, \qquad c_1^4 \geq c_1^2, \qquad d_1^4 \geq d_1^2, \tag{6.336}$$

and

$$\begin{aligned} 2a_1b_1c_1d_1 &\geq a_1b_1 + c_1d_1, \\ 2a_1b_1c_1d_1 &\geq a_1d_1 + c_1b_1, \\ 2a_1b_1c_1d_1 &\geq a_1c_1 + b_1d_1. \end{aligned} \tag{6.337}$$

In addition, it follows that

$$a_1^8 + b_1^8 + c_1^8 + d_1^8 \geq 4\sqrt[4]{a_1^8 \cdot b_1^8 \cdot c_1^8 \cdot d_1^8}. \tag{6.338}$$

In order to prove inequality (6.330) it suffices to prove that

$$\begin{aligned} 3\left(a_1^2 + b_1^2 + c_1^2 + d_1^2\right) \geq (a_1 + b_1)(c_1 + d_1) + (a_1 + d_1)(b_1 + c_1) \\ + (a_1 + c_1)(b_1 + d_1). \end{aligned} \tag{6.339}$$

However, we have

$$\begin{aligned} &a_1^2 + b_1^2 + c_1^2 + d_1^2 \geq a_1c_1 + a_1d_1 + b_1c_1 + b_1d_1 = (a_1 + b_1)(c_1 + d_1), \\ &a_1^2 + b_1^2 + c_1^2 + d_1^2 \geq (a_1 + c_1)(b_1 + d_1), \\ &a_1^2 + b_1^2 + c_1^2 + d_1^2 \geq (a_1 + d_1)(b_1 + c_1). \end{aligned} \tag{6.340}$$

Adding the above inequalities by parts, we deduce (6.330). □

Remark 6.8 The existence of at least one figure satisfying the requirements of the problem is a consequence of the following reasoning:

Consider the circle C_1 with center I and radius $r + 2\sqrt{2}$, as well as the circle C_2 with center I and radius $2r + \sqrt{2}$. If O is an arbitrary point of C_1 such that $(IO) = r$, then $(AO) \geq AS$ with $(AS) = (AI) - (IS)$. Thus

$$(AI) \geq 2r + \sqrt{2}.$$

Therefore,

$$(AS) \geq 2r + \sqrt{2} - r - \frac{\sqrt{2}}{2},$$

which implies that

$$(AS) \geq r + \frac{\sqrt{2}}{2},$$

and thus

$$(AO) \geq (IO) + \frac{\sqrt{2}}{2}.$$

Similarly, one can prove that

$$(BO) \geq (IO) + \frac{\sqrt{2}}{2},$$

$$(CO) \geq (IO) + \frac{\sqrt{2}}{2},$$

$$(DO) \geq (IO) + \frac{\sqrt{2}}{2}.$$

Remark 6.9 If $a = b = c = d = \frac{1}{2}$, it follows that the inequality does not hold.

6.3.12 Let a circle (O, R) be given and let a point A be on this circle. Consider successively the arcs AB, BD, DC such that

$$\text{arc}\, AB < \text{arc}\, AD < \text{arc}\, AC < 2\pi.$$

Using the center K of the arc BD, the center L of BD, and the corresponding radii, we draw circles that intersect the straight semilines AB, AC at the points Z and E, respectively (see Fig. 6.64). If

$$A' \equiv AL \cap DC, \qquad K' \equiv AK \cap BD,$$

prove that

$$\frac{3}{4}(AB \cdot AZ + AC \cdot AE) < 2R^2 + \frac{R(AK' + AL')}{2} + \frac{AB^2 + AC^2}{4}. \tag{6.341}$$

Is this inequality the best possible?

Solution We are going to use the following

Lemma 6.13 *Let the circle (O, R) be given and its points A, K. With center the point K and radius smaller that the length of the chord AK we draw a circle that intersects the initial circle at the points B and D. If Z is the intersection point of*

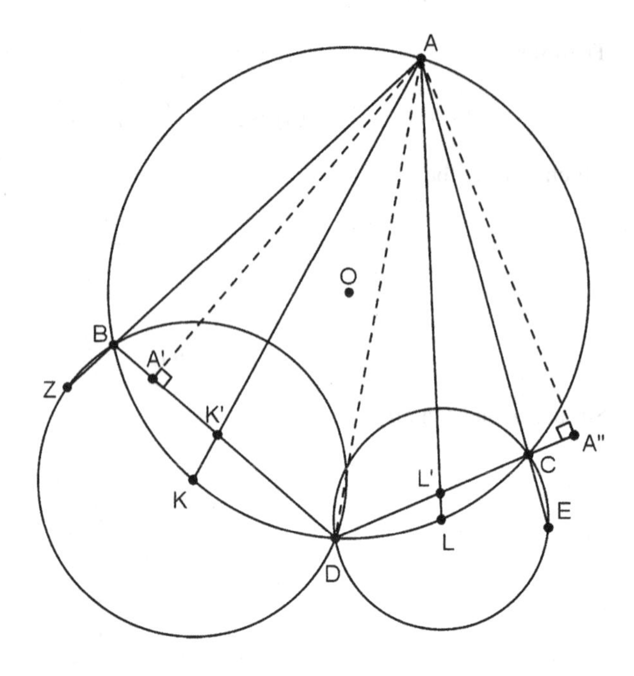

Fig. 6.64 Illustration of Problem 6.3.12

the straight semiline AB with the circle (K, KD) *and H the common point of the straight semiline AD with the circle* (K, KD), *then*

$$AZ = AD \quad \textit{and} \quad AH = AB.$$

Indeed, since (see Fig. 6.65)

$$\widehat{ADK} + \widehat{KBA} = \pi,$$

it follows that

$$\widehat{ADK} + \widehat{KZB} = \pi,$$

and thus

$$\widehat{ADK} = \widehat{AZK} \quad \text{with } \widehat{ZAK} = \widehat{KAD}.$$

Hence, the triangles AZK and ADK are equal. Therefore, $AZ = AD$. Similarly, the triangles ABK and AHK are equal, and thus $AH = AB$. The assertion of the lemma follows. □

Back to the original problem, let A' be a point of the straight line BD and A'' a point of DC. It holds

$$AA' \perp BD \quad \text{and} \quad AA'' \perp DC,$$

Fig. 6.65 Illustration of Problem 6.3.12

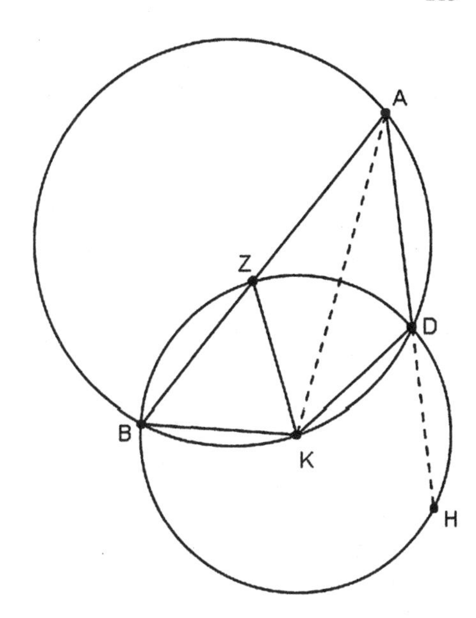

and therefore

$$AA' \le AK' \quad \text{and} \quad AA'' \le AL'. \tag{6.342}$$

In order to prove (6.341), it should be enough to verify the relation

$$0 < 8R^2 - 2AB \cdot AD - 2AC \cdot AD + AB^2 + AC^2, \tag{6.343}$$

or equivalently,

$$0 < 8R^2 - 2AB \cdot AD - 2AC \cdot AD + AB^2 + AC^2 + 2AD^2 - 2AD^2. \tag{6.344}$$

Using (6.342) and the fact that in a triangle ABC the relation

$$bc = 2Rh$$

holds true, where R is the radius of the circumcircle and h the height drawn from the vertex A, it follows that (6.344) yields

$$0 < 2\big((2R)^2 - AD^2\big) + (AB - AD)^2 + (AC - AD)^2, \tag{6.345}$$

which holds true and this completes the proof of the inequality (6.341).

Remark 6.10 Because of the compactness, the inequality (6.341) cannot be improved, otherwise AB and AC would be identical and simultaneously would coincide with AD. □

Appendix

And since geometry is the right foundation of all painting, I have decided to teach its rudiments and principles to all youngsters eager for art.

Albrecht Dürer (1471–1528)

A.1 The Golden Section

Dirk Jan Struik (1894–2000), Former Professor of Mathematics, Massachusetts Institute of Technology, USA[1]

A good mathematician, it has been said, must also be something of an artist. He studies his field, Henri Poincaré, the great French mathematician, has said, not because it is useful, but because it is beautiful. Whatever truth there may be in such statements, it is certain that there always have been many connections between mathematicians and the arts, especially connections with music, architecture and painting, often based on philosophical considerations as those of Pythagoras and Plato in Antiquity. Many well-known mathematicians, from Euclid in classical days to Euler and Sylvester in more recent times, have shown profound interest in music, an interest also shared by modern mathematicians.

It is not that mathematicians are more likely to be good piano, cello or flute players than physicians, lawyers or undertakers. It is the theory of music that, ever since the days of Pythagoras, has drawn mathematical attention to the different harmonics in the scale and their quantitative relationship. However, in this article, we shall not deal with this aspect of the relationship of mathematics and the arts, but with another such relationship, in which architecture and painting are involved. This relationship is known as *the division of a line segment in extreme and mean ratio*, a term we find in Euclid's "Elements", written ca 300 BC in Alexandria, the new city on the Nile Delta founded by Alexander the Great (the Greek term is: *άκρος και μέσος λόγος*).

[1]Reprinted from *Mathematics in education* (ed. Th.M. Rassias), University of LaVerne Press, California, 1992, pp. 123–131 with the kind permission of the editor.

S.E. Louridas, M.Th. Rassias, *Problem-Solving and Selected Topics in Euclidean Geometry*, DOI 10.1007/978-1-4614-7273-5,

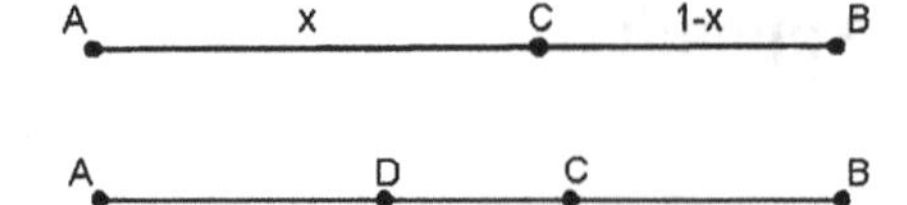

Fig. A.1 Golden Section

Fig. A.2 Golden Section

Euclid gives two constructions for it and uses it repeatedly in the *Elements*, especially in order to construct a regular pentagon and a regular decagon inscribed in a circle, this again in connection with the five regular solids, the so-called Platonic bodies, and especially with the regular dodecahedron and the regular icosahedron. Plato has explained how this extreme and mean ratio could be connected with philosophical problems, and especially with Plato's cosmogony, in which the regular bodies play a fundamental role. The cosmic role of the ratio (or *section* ($\tau o\mu\acute{\eta}$), as it is sometimes called) made mathematicians in Renaissance days call it *Golden Section* and even *Divine Proportion*. We shall occasionally denote it by **G.S.**[2] The ratio, **G.S.**, is obtained, as Euclid explains, by taking a line segment AB (Fig. A.1) and finding a point C between A and B such that ($AC > CB$):

$$\frac{AC}{CB} = \frac{AB}{AC} \tag{A.1}$$

in words: the longest part is to the smallest part as the whole segment is to the longest part. There exists, of course, also another point D between A and B which determines a **G.S.**, but then $AD < DB$, see Fig. A.2. In order to understand better why the **G.S.** has interested, even excited, so many persons throughout the ages, mathematicians as well as artists (and even mystics), let us start with the Pythagoreans, a philosophical sect in ancient Greece, flowering between ca 500–250 BC and dating their origin to the sage Pythagoras, mathematician and student of the universe. Members of this sect believed strongly in the mathematical symbolism both for scientific and social–ethical reasons. A favorite symbol was the five pointed star called *pentagram* (Fig. A.3) obtained by taking a regular pentagon $ABCDE$ and extending its sides to their intersections $PQRST$. You can see it also as an overlapping of five letters A, Greek alpha. Hence the name *pentalpha* (Fig. A.4) for this figure, which, incidentally, can be drawn in one stretch without lifting the pencil from the paper. Why the pentalpha had assumed this favorite, even magical, character is not quite clear, but it was a figure of interest already long before the Pythagoreans made

[2]The literature on the Golden Section is large and of varied character. Useful of older literature is R.C. Archibald, *Golden section*, The American Mathematical Monthly **25** (1918) 232–235, who cites Emma C. Ackermann, The American Mathematical Monthly **2** (1985) 260–264, who wrote this account based on F.C. Pfeiffer, *Der Goldene Schnitt*, Augsburg, 1885. A newer account with many details in H.E. Huntley, *The divine proportion*, Dover, New York, 1970, VII + 181 pp. a book that calls itself *A study in mathematical beauty*. See also H.S.M. Coxeter, *Introduction to geometry*, Wiley, New York/London, 1961, XIV + 443 pp., esp. Chap. 11. A recent, quite technical, work is R. Herz-Fischer, *A mathematical history of division in extreme and mean ratio*, University Press, Waterloo, Ontario, 1987, XVI + 191 pp. Euclid's *Elements* can be studied in the English version, with ample commentary, by T.L. Health, *The thirteen books of Euclid's elements*, Cambridge University Press, 1956, Dover reprint, 3 volumes.

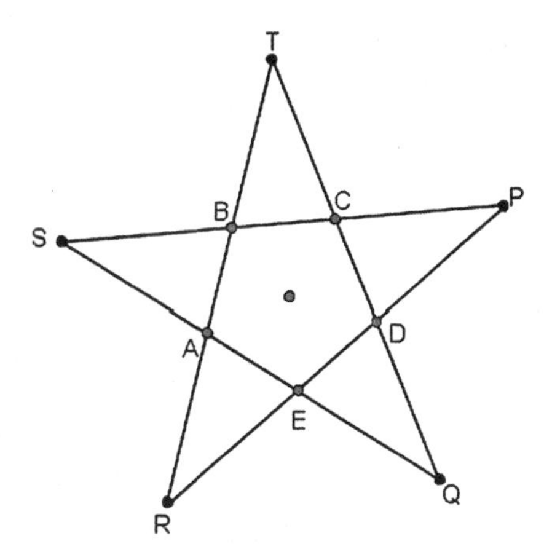

Fig. A.3 Pentagram

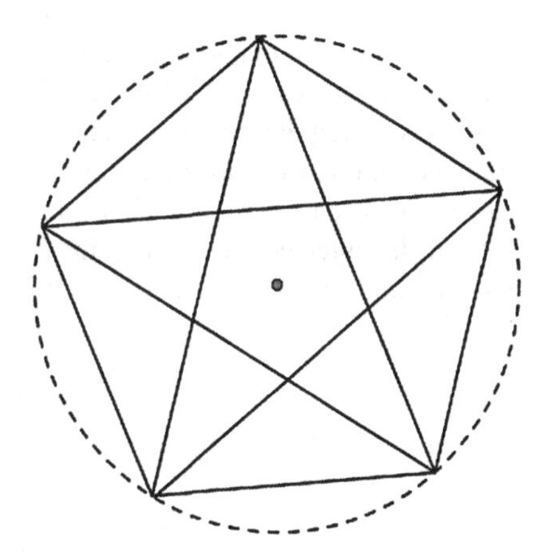

Fig. A.4 Pentalpha

it a subject of philosophical, even mathematical, interest. We find it, for instance, on ancient Babylonian drawings, and, for all we know, it may date back to the Stone Age. Did its likeness to the twinkling stars in heaven have something to do with it? For the Pythagoreans it was a symbol of health and of recognition; when you saw a pentagram on a house you could expect hospitality and friendship. Later, in the European Middle Ages and later, it served as an apotropaion, a means to ward off danger, or evil. In Central Europe, it was supposed to guard against a female spirit called *Drude*, hence its name *Drudenfuss* (Drude's feet). Doctor Faust, Goethe's drama, had such a figure on the door step of his study but the devil in the shape of Mephistopheles was still able to trespass because the top of the *Drude's foot* pointed outward was not quite closed.[3]

[3]The text of Goethe's *Faust* says: Der Drudenfuss auf Eurer Schwelle ...
Beschaut es recht! es ist nicht gut gezogen ...
(The Drude's foot on your doorstep ... look carefully, it is not drawn correctly.)

Fig. A.5 Regular Pentagon

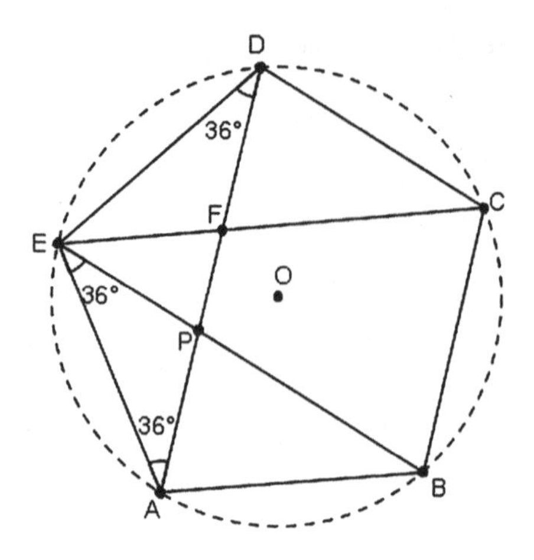

So much for the magical properties of the pentagram. It has also interesting mathematical properties, as the Pythagoreans, and Euclid with them, were well aware. Let us take a regular pentagon. The diagonals of this pentagon form again a pentagram and in this we can again find a regular pentagon, and so forth.

Moreover, any two diagonals intersect in a **G.S.** Take, for instance, diagonals AD and BE, intersecting at the point P (Fig. A.5). Triangles PEA and EDF are both isosceles and, since their angles are 36°, 72°, and 72°, are similar. Hence ($ED = PD$)

$$\frac{AD}{AE} = \frac{AE}{AP},$$

or

$$\frac{AD}{PD} = \frac{PD}{AP} = \tau.$$

We designate this ratio by τ (some use the letter ϕ or e), taking $AD = 1$.

If we have a close look at the triangle DEP, isosceles with top angle at D of 36°, and angles at E and P of 72°, and bisect the angle at E, the bisector hitting DP at F (Fig. A.6), then we see that the triangle FDE is also isosceles. If we take $EP = 1$, then $ED = \tau$, hence F divides DP in the **G.S.** Here $DF = 1$, $DFP = \tau$ and $\tau = 2\cos 36° = 1.618033989\ldots$.

Euclid studies this triangle in Book *IV*, Prop. 10 and uses it to construct a regular pentagon in a circle. And since the angle at D is 36°, EP is the side of the regular decagon (polygon of 10 sides) inscribed in a circle with center D and radius $DE = DP$. This gives us the possibility of constructing a regular decagon, and hence also a regular pentagon, in a circle as soon as we know how to divide a line segment in extreme and mean ratio. Euclid gives two constructions for this purpose, one in Book *II* (on area), the other in Book *VI* (on propositions). We replace them by the construction of Fig. A.7. Let AB be the line segment to be divided into extreme and mean ratio. Take $GB = AB$ perpendicular to AB at the point B and let GB be the diameter of the circle with center C halfway between B and G. Then connect A

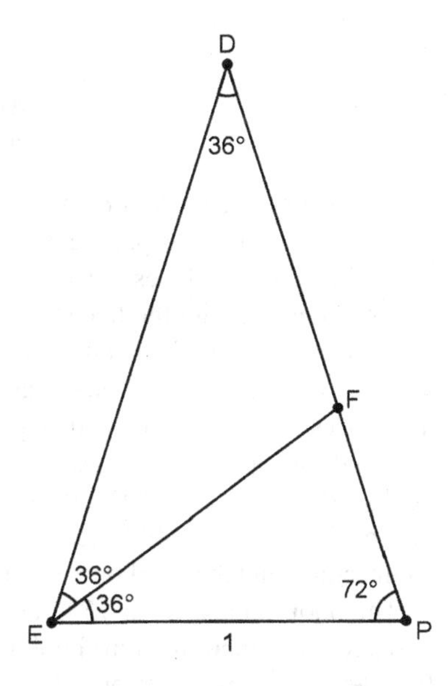

Fig. A.6 Isosceles triangle with angles 36°, 72°, 72°

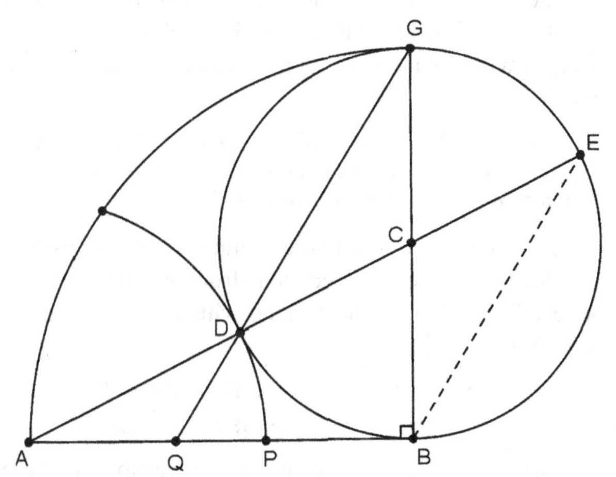

Fig. A.7 Construction of the Golden Section

with C. This line intersects the circle at D (and, continued, also at E). Then, when the circle with radius AD and center A intersects AB in P, this P provides on AB the desired ratio (Fig. A.7). Indeed, since $DE = AB$, $AD = AP$, we can write

$$\begin{aligned} AB^2 &= AD \times AE = AD(AD + AB) \\ &= AD^2 + AD \times AB \\ &= AP^2 + AP \times AB. \end{aligned} \tag{A.2}$$

Hence

$$AB(AB - AP) = AP^2 = AB \times PB,$$

or

$$\frac{AP}{PB} = \frac{AB}{AP} = \tau. \tag{A.3}$$

When GD intersects AB in Q, it also divides AB in the Golden Section. This follows from the fact that BE is parallel to GD.

Euclid also introduces our ratio in the first propositions opening his Book *XIII*, the book dealing with the five regular (Platonic) bodies, the regular tetrahedron, the hexahedron or cube, the octahedron, the dodecahedron, and the icosahedron. These last two solids have particular relations with the extreme and mean ratio because their faces are related to regular pentagons. This was recognized throughout the ages and especially in Renaissance days, when the Franciscan monk Luca Paccioli published a book called *Divina Proportione*, a book in three parts, of which the first one, written in 1497, deals with the Golden Section, the second book with architecture, and the third one with the regular solids.[4]

The book published in 1509 and republished in 1956 has pictures ascribed to Leonardo da Vinci; the third book is based on a text by the painter–mathematician Pier della Francesca. Among those great men of the Renaissance who also were deeply moved by the mathematical and philosophical attraction of the Platonic bodies and, with them the Golden Section, was Kepler. In an often quoted passage, he claimed:

> *Geometry has two great treasures, one is the theorem of Pythagoras, the other the division of a line into extreme and mean ratio. The first we may compare to a measure of gold, the second we may name a precious jewel.*[5]

Jewels have considerable esthetic appeal. The esthetic value of the Golden Section has often been appreciated, from Antiquity to the present time. It has been believed that a rectangle, formed with sides in Golden Section relationship, hence (Fig. A.7)

$$\frac{AB}{BC} = \frac{AB + BC}{AB},$$

or, when $BC = 1$, $AB = \tau$, is more agreeable to the eye than any other type of rectangle. We find this shape in buildings, for instance, those of Antiquity like the Parthenon and those constructed under Greek inspiration (Fig. A.8); it is also taken seriously by some modern architects like Le Corbusier.

The Leipzig psychologist, Gustav Theodor Fechner, experimented in the 1870s with a large number of persons who were asked which type of rectangular frame was most pleasing to their way of thinking, and it turned out that the $1 : \tau$ frame was statistically the winner. This was an application of Fechner's *psychophysics*, namely

[4] Luca Paccioli, *Divina Proportione*, Venice, 1509, republished in Verona 1956. German translation by C. Winterberg, Vienna, 1889, 1896. Paccioli must have met Leonardo da Vinci at the Milan court of Ludovico Sforza, to whom his book is dedicated.

[5] See Archibald, note 1, footnote 2, p. 234.

Fig. A.8 Parthenon

experimental esthetics. It involved questions about the best shapes of windows, picture frames, book forms, playing cards, even snuff boxes.[6] The same ratio has also been found pleasing in human and animal bodies, as well in morphology, in general.

So far we have discussed the Golden Section mainly from a geometrical point of view, following the ancient method of the Greeks. Let us now introduce some algebra, the type of mathematics introduced and developed in Europe during the late Middle Ages and Renaissance days under the influence of Islamic mathematics (as the name *algebra*, derived from the Arabic, indicates).

Let us take (Fig. A.1) a straight line segment $AB = 1$, then take $AC = x$, $CB = 1 - x$, $x > 1 - x$. Then

$$\frac{x}{1-x} = \frac{1}{x} = \tau, \tag{A.4}$$

or

$$x^2 + x - 1 = 0 \quad \text{and} \quad \tau^2 - \tau - 1 = 0. \tag{A.5}$$

Hence

$$x = \frac{\sqrt{5}-1}{2}, \quad 1 - x = \frac{3-\sqrt{5}}{2}, \tag{A.6}$$

and

$$\tau = \frac{\sqrt{5}+1}{2}, \quad \frac{1}{\tau} = \frac{\sqrt{5}-1}{2} = x. \tag{A.7}$$

We see that

$$-\frac{1}{\tau} = -x$$

[6] G.F. Fechner, *Vorschule der Aesthetik*, 1876. Some people prefer the ratio $\frac{1}{\sqrt{2}}$, that of the side to the diagonal of a square. Hence about $10/14$ instead of about $10/16$.

is the other root of the equation with respect to x. We find this value of τ already in Euclid, but in geometrical form (Book *XIII*, Prop. 1).

We conclude that

$$\tau = 1.6180339\ldots, \quad \frac{1}{\tau} = 0.6180339\ldots = x. \tag{A.8}$$

We saw already that

$$\tau = 2\cos 36^\circ.$$

This number τ has many interesting properties, due to Eq. (A.7).

With it we can form a geometrical series

$$1 + \tau + \tau^2 + \tau^3 + \cdots + \tau^n + \cdots \tag{A.9}$$

and replace τ^2 with $\tau + 1$; we obtain

$$\begin{aligned} \tau^3 &= \tau(\tau+1) = \tau^2 + \tau = 1 + 2\tau, \\ \tau^4 &= \tau(1+2\tau) = 1 + 3\tau, \\ &\vdots \end{aligned} \tag{A.10}$$

Hence the series (A.9) can also be written as

$$(1+\tau) + (1+2\tau) + (1+3\tau) + \cdots + (1+n\tau) + \cdots \tag{A.11}$$

which is an *arithmetical* series. The same holds for $-1/\tau$.

A second, even more interesting property can be observed when we connect τ with the theory of continued fractions, a theory also dating from Renaissance days, where we find a book on the subject by P.A. Cataldi (1613). Then, in a book by A. Girard of 1634 we find a reasoning equivalent to the following:

$$\begin{aligned} \tau &= 1 + \frac{1}{\tau} \\ &= 1 + \frac{1}{1 + \frac{1}{\tau}} \\ &= 1 + \frac{1}{1 + \frac{1}{1 + \frac{1}{\tau}}}, \quad \text{etc.,} \end{aligned} \tag{A.12}$$

which gives us τ in the form of a continued fraction:

$$\tau = 1 + \frac{1}{1 + \frac{1}{1 + \frac{1}{\tau}}}, \quad \text{etc.} \tag{A.13}$$

Such a continued fraction has partial fractions (convergents), such as

$$1, \quad 1+\frac{1}{1}=2, \quad 1+\frac{1}{1+\frac{1}{1}}=\frac{3}{2}, \quad 1+\frac{1}{1+\frac{1}{1+\frac{1}{1}}}=\frac{5}{3}, \quad \text{etc.}$$

We thus obtain the sequence

$$1,\ 2,\ \frac{3}{2},\ \frac{5}{3},\ \frac{8}{5},\ \frac{13}{8},\ \frac{21}{13},\ \frac{34}{21},\ \frac{55}{34},\ \frac{89}{55},\ \frac{144}{89},\ldots \tag{A.14}$$

that is,

$$1,\ 2,\ 1.5,\ 1.66,\ 1.60,\ 1.625,\ 1.6154,\ 1.6190,\ 1.6176,\ldots$$

which is a sequence of numbers oscillating around

$$\tau = 1.618033989\ldots,$$

coming closer and closer to τ, the difference between them and τ becoming smaller than any given small number δ, so that τ is the limit (we omit here the exact proof).

The numerators and denominators of the ratios of the sequence (A.14) (we add 1 in front) are of the form

$$1, 1, 2, 3, 5, 8, 13, 21, 34, 55, 89, 144, 233, 377, \ldots \tag{A.15}$$

and are such that each one of them is the sum of the two preceding numbers. If we write the sequence (A.15) in the form

$$u_1, u_2, u_3, u_4, \ldots, u_n, \ldots, \tag{A.16}$$

we have

$$u_1 = u_2 = 1, \quad u_3 = 2, \quad u_4 = 3, \quad \text{etc.}$$

Then the following recursive relation holds true

$$u_n = u_{n-1} + u_{n-2} \tag{A.17}$$

and the fraction

$$\frac{u_{n+1}}{u_n}$$

approaches τ as n increases. This sequence is called a *Fibonacci set*, after the merchant–mathematician Leonardo of Pisa, also called Fibonacci (member of the house of Bonacci). This merchant, on his many travels, picked up much mathematics in Islamic countries, which inspired him to write a book called *Liber Abaci* (1202), the first important text on Arabic mathematics in Latin Europe.[7] It has many

[7] The *Liber Abaci* was published for the first time in 1857 by Prince B. Boncompagni in Rome. The rabbit problem can be found in pp. 283–284. See, e.g., R.C. Archibald, The American Mathematical Monthly **25** (1918) 235–238.

problems with solutions, all in the new at that time decimal position system (the so-called *Hindu–Arabic number system*). One of the problems is the following:

> *A man has a pair of rabbits. We wish to know how many pairs can be bred from it in one year, if the nature of these rabbits is such that they breed every month one other pair and begin to breed in the second month after their birth.*

Fibonacci then finds: at the beginning 1 pair, after first month 2, after the third month 3, after the third month 5, etc., after the twelfth month 377. This set is a Fibonacci set.

These numbers have many interesting properties. For example, there is the equation

$$u_{n-1}u_{n+1} - u_n^2 = (-1)^n,$$

found by the Scottish mathematician Robert Simpson[8] for several years in a paper of 1753 dealing with Girard's remarks of 1634, and the equation

$$2^n\sqrt{5}\,u_n = (1+\sqrt{5})^n - (1-\sqrt{5})^n$$

found by J. P. M. Binet[9] in a memoir on linear difference equations, and useful in showing that

$$\frac{u_{n+1}}{u_n}$$

for growing n tends towards τ. Several mathematicians have been—and are—interested in these numbers that they have been published in *The Fibonacci Quarterly*.

Another case, in which Fibonacci numbers play a role, is that of phyllotaxis, from phyllon ($\phi\acute{\upsilon}\lambda\lambda o\nu$), leaf, and taxis ($\tau\acute{\alpha}\xi\iota s$), arrangement. This is the field that deals with the way leaves are placed around the stems (or twigs) of plants. It is old, having had the attention of Greek and Renaissance students as Leonard Fuchs (1452), after whom the Fuchsia is named. Linnaeus, in the eighteenth century, paid also attention to this arrangement. But in the 1830s, two German botanists, Karl Schimper and Alexander Braun, influenced by Pythagorean inspired Naturphilosophie of the Jena professor Lorenz Oken, found out that growth of the leaves in the stem has a forward direction in a spiral such that the leaves are arranged in regular cyclic mathematical patterns, each species having its own. The number of leaves along the spiral (or helix) and the number n of rotations of this spiral between two leaves that are precisely above each other determines the arrangement of the leaves. If in the n rotations we meet k leaves, then we speak of an (n, m) phyllotaxis. With, for instance, a beech we have $(1, 3)$, for an apricot $(2, 5)$, a pear $(3, 8)$ phyllotaxis.

---

[8]R. Simson, *Philosophical* Transactions, 1753.

[9]J.P.M. Binet, Comptes Rendus Académie Française **17** (1843) 563.

Schimper found out that these numbers (n, m) form a Fibonacci set:

$$1, 2, 3, 5, 8, \ldots, \quad \text{etc.}$$

There are, of course, irregularities, but the rule stands in most cases. Larger numbers of the Fibonacci set also occur. The arrangement of florets in a sunflower, on 21 clockwise, 34 counterclockwise spirals, is an example. Another case is that of the scales of a pineapple.[10] There are also relations of the **G.S.** and the Fibonacci numbers with the logarithmic spiral, and this again with the shells of a large number of living creatures, from the very small foraminifera to such a well-known beauty as the *chambered* nautilus, of the Indo–Pacific ocean. For this and other applications, we can refer to Coxeter and D'Arcy Thompson.[11]

[10]Oken, in his turn, was influenced by the *Naturphilosophie* of Schelling. On phyllotaxis see further the books mentioned in note 2 by Coxeter, pp. 169–172 and Huntley, pp. 161–164. For the Schimper–Braun contribution, see A.A. Braun, Dictionary Scientific Biography **2** (1970) 426.

[11]H.S.M. Coxeter, *The golden section, phyllotaxis and Wijthoff's game*, Scripta Mathematica **19** (1953) 139.

D'Arcy Thompson, *On growth and form*, Cambridge University Press, 1917, 2nd ed., 1942, pp. 912–933.

N. N. Vorob'ev, *Fibonacci numbers*, transl. by H. More, Blaisdell, New York, London, 1961.

References

1. Alexanderson, G. L., Klosinski, L. F., & Larson, L. C. (Eds.) (1985). *The William Lowell Putnam mathematical competition. Problems and solutions, 1965–1984*. Washington: Math. Assoc. of America.
2. Alexandrov, A. D. (1945). The isoperimetric problem. *Doklady Akademii Nauk SSSR, 50*, 31–34.
3. Alexandrov, A. D., Kolmogorov, A. N., & Lavrent'ev, M. A. (1999) *Mathematics–its contents, methods and meaning*. Mineola: Dover. (Translation edited by S. H. Gould.)
4. Altshiller-Court, N. (1952). *College geometry: an introduction to the modern geometry of the triangle and the circle*. New York: Dover.
5. Almgren, F. J. (1964). An isoperimetric inequality. *Proceedings of the American Mathematical Society, 15*, 284–285.
6. Andreescu, T. (2010). Problem J149. *Mathematical Reflections, 1*, 1.
7. Andreescu, T. (2011). Problem J193. *Mathematical Reflections, 3*, 1.
8. Andreescu, T., Andrica, D., & Barbu, C. (2011). Problem O196. *Mathematical Reflections, 3*, 4.
9. Andreescu, T., & Gelca, R. (2001). *Mathematical olympiad challenges*. Boston: Birkhäuser.
10. Andreescu, T. & Feng, Z. (Eds.) (2002). *Mathematical olympiads. Problems and solutions from around the world, 1999–2000*. Washington: Math. Assoc. of America.
11. Andreescu, T., Feng, Z., & Lee, G. Jr. (Eds.) (2003). *Mathematical olympiads. Problems and solutions from around the world, 2000–2001*, Washington: Math. Assoc. of America.
12. Andreescu, T., & Feng, Z. (2002). *102 combinatorial problems*. Boston: Birkhäuser.
13. Andreescu, T., & Kedlaya, K. (1998). *Mathematical contests 1996–1997, olympiads problems and solutions from around the world*.
14. Andrica, D., & Lu, K. (2008). Problem J77. *Mathematical Reflections, 2*, 12.
15. Andrica, D. (2009). Problem S128. *Mathematical Reflections, 4*, 10.
16. Andrica, D. (2010). Problem O177. *Mathematical Reflections, 6*, 30.
17. Andrica, D. (2010). Problem J160. *Mathematical Reflections, 3*, 4.
18. Andrica, D. (2010). Problem O165. *Mathematical Reflections, 4*, 28.
19. Andrica, D. (2010). Problem O179. *Mathematical Reflections, 6*, 32.
20. Andrica, D. (2010). Problem S167. *Mathematical Reflections, 4*, 15.
21. Andreescu, T., & Andrica, D. (2006). *Complex numbers from A to … Z*. Boston: Birkhäuser.
22. Baker, H. F. (1963). *An introduction to plane geometry*. London: Cambridge University Press.
23. Baleanu, A. R. (2009). Problem J125. *Mathematical Reflections, 4*, 1.
24. Barbu, C. (2010). Problem O147. *Mathematical Reflections, 1*, 5.
25. Becheanu, M. (2011). Problem O196. *Mathematical Reflections, 3*, 4.
26. Borsenko, I. (2010). Problem J146. *Mathematical Reflections, 1*, 1.
27. Beckenbach, E. F., & Bellman, R. (1961). *Inequalities*. New York: Springer.

S.E. Louridas, M.Th. Rassias, *Problem-Solving and Selected Topics in Euclidean Geometry*, DOI 10.1007/978-1-4614-7273-5,

28. Benson, D. G. (1970). Sharpened forms of the plane isoperimetric inequality. *The American Mathematical Monthly*, *77*, 29–31.
29. Blaschke, W. (1915). *Kreis und Kugel*. (2. Aufl. de Gruyter, Berlin, 1956.)
30. Brand, L. (1944). The eight-point circle and the nine-point circle. *The American Mathematical Monthly*, *51*, 84–85.
31. Burago, Yu. D., & Zalgaller, V. A. (1988). *Geometric inequalities*. Berlin: Springer. (Translated from Russian by A. B. Sossinsky.)
32. Busemann, H. (1958). *Convex surfaces*. New York: Interscience.
33. Busemann, H. (1960). Volumes and areas of cross-sections. *The American Mathematical Monthly*, *67*, 248–250.
34. Cabrera, R. B. (2009). Problem S124. *Mathematical Reflections*, *4*, 2.
35. Cabrera, R. B. (2009). Problem U123. *Mathematical Reflections*, *4*, 3.
36. Cesari, L. (1956). *Surface area*. Princeton: Princeton University Press.
37. Coolidge, J. L. (1939). A historically interesting formula for the area of a quadrilateral. *The American Mathematical Monthly*, *46*, 345–347.
38. Coxeter, H. S. M. (1969). *Introduction to geometry*. New York: Wiley.
39. Coxeter, H. S. M., & Greitzer, S. L. (1967). *Geometry revisited*. New York: Random House.
40. Djukic, D., Jankovic, V., Matic, I., & Petrovic, N. (2006). *The IMO compendium. A collection of problems suggested for the international mathematical olympiads: 1959–2004*. New York: Springer.
41. Enescu, B. (2010). Problem J174. *Mathematical Reflections*, *5*, 1.
42. Fleming, R. J., & Jamison, J. E. (2003). *Isometries in Banach spaces: function spaces*. Boca Raton: Chapman & Hall/CRC.
43. Gallanty, W. (1913). *The modern geometry of the triangle* (2nd ed.). London: Hodgson.
44. Gleason, A. M., Greenwood, R. E., & Kelly, L. M. (1980). *The William Lowell Putnam mathematical competition, problems and solutions: 1938–1964*. Washington: Math. Assoc. of America.
45. Gowers, T. (Ed.) (2008). *The Princeton companion to mathematics*. Princeton, Oxford: Princeton University Press.
46. Hahn, L. (1960). *Complex numbers and geometry*. New York: Math. Assoc. of America.
47. Hardy, G. H., Littlewood, J. E., & Pólya, G. (1952). *Inequalities* (2nd ed.). Cambridge: Cambridge University Press.
48. Hardy, G. H. (1975). *A course of pure mathematics* (10th ed.). Cambridge: Cambridge University Press.
49. Heath, T. L. (1956). *The thirteen books of Euclid's elements (3 volumes)*. New York: Dover.
50. Hilbert, D., & Cohn-Vossen, S. (1952). *Geometry and the imagination*. New York: Chelsea
51. Honsberger, R. (1995). *Episodes in nineteenth and twentieth century Euclidean geometry*. Washington: Math. Assoc. of America.
52. Honsberger, R. (1996). *From Erdős to Kiev. Problems of olympiad caliber*. Washington: Math. Assoc. of America.
53. Jankovic, V., Kadelburg, Z., & Mladenovic, P. (1996). *International and Balkan Mathematical Olympiads 1984–1995*. Beograd: Mathematical Society of Serbia (in Serbian).
54. Johnson, R. A. (1929). *Modern geometry: an elementary treatise on the geometry of the triangle and the circle*. Boston: Houghton Mifflin.
55. Johnson, R. A. (1960). *Advanced Euclidean geometry*. New York: Dover.
56. Kay, D. C. (2001). *College geometry. A discovery approach* (2nd ed.). Boston: Addison Wesley.
57. Kazarinoff, N. D. (1975). *Geometric inequalities*. Washington: Math. Assoc. of America.
58. Kedlaya, K. S., Poonen, B., & Vakil, R. (2002). *The William Lowell Putnam mathematical competition 1985–2000, problems, solutions and commentary*. Washington: Math. Assoc. of America.
59. Kimberling, C. (1998). Triangle centers and central triangles. *Congressus Numerantium*, *129*, 1–295.

60. Klamkin, M. S. (1986). *International mathematical olympiads 1979–1985 and forty supplementary problems*. Washington: Math. Assoc. of America.
61. Klamkin, M. S. (1988). *International mathematical olympiads 1979–1986*. Washington: Math. Assoc. of America.
62. Klee, V., & Wagon, S. (1991). *Old and new unsolved problems in plane geometry and number theory* (rev. ed.). Washington: Math. Assoc. of America.
63. Klein, F. (1980). Famous problems of elementary geometry: the duplication of the cube, the trisection of the angle, and the quadrature of the circle. In *Famous problems and other monographs* (pp. 1–92). New York: Chelsea.
64. Kupetov, A., & Rubanov, A. (1975). *Problems in geometry*. Moscow: MIR.
65. Lachlan, R. (1893). *An elementary treatise on modern pure geometry*. London: Macmillan.
66. Lang, S. (1985). *The beauty of doing mathematics. Three public dialogues*. New York: Springer.
67. Larson, L. C. (1983). *Problem-solving through problems*. New York: Springer.
68. Lovasz, L. (1979). *Combinatorial problems and exercises*. Amsterdam: North-Holland.
69. Louridas, S. E. (2000–2001). *Euclide B'. Hellenic Mathematical Society* (Vol. 34).
70. Lozansky, E., & Rousseau, C. (1996). *Winning solutions*. New York: Springer.
71. Mauldin, R. D. (Ed.) (1981). *The Scottish book. Mathematics from the Scottish café*. Boston: Birkhäuser.
72. Mitrinovic, D. S. (1970). *Analytic inequalities*. Heidelberg: Springer.
73. Mitrinovic, D. S., Pecaric, J. E., & Volonec, V. (1989). *Recent advances in geometric inequalities*. Dordrecht: Kluwer Academic.
74. Modenov, P. S. (1981). *Problems in geometry*. Moscow: MIR.
75. Modenov, P. S., & Parhomenko, A. S. (1965). *Geometric transformations*. New York: Academic Press.
76. Ogilvy, C. S. (1969). *Excursions in geometry*. New York: Dover.
77. Osserman, R. (1978). The isoperimetric inequality. *Bulletin of the American Mathematical Society, 84*, 1182–1238.
78. Pedoe, D. (1995). *Circles: a mathematical view*. Washington: Math. Assoc. of America.
79. Penrose, R. (2004). *The road to reality—a complete guide to the laws of the universe*. London: Jonathan Cape.
80. Pólya, G. (1954). *Mathematics and plausible reasoning*. Princeton: University Press.
81. Pólya, G. (1957). *How to solve it* (2nd ed.). Princeton: Princeton University Press.
82. Pólya, G., & Szegö, G. (1978). *Problems and theorems in analysis* (Vol. 1). Berlin: Springer.
83. Posamentier, A. S., & Salkind, C. T. (1997). *Challenging problems in geometry*. New York: Dover.
84. Pop, O. T., Minculete, N., & Bencze, M. (2013). *An introduction to quadrilateral geometry*. Bucuresti: Editura Didactica si Pedagocica.
85. Prasolov, V. V. (1986). *Problems of plane geometry* (Vols. 1 and 2). Moscow: Nauka.
86. Rassias, M. Th. (2011). Problem-solving and selected topics in number theory. In *The spirit of the mathematical olympiads*. New York: Springer. (Foreword by Preda Mihailescu.)
87. Russell, B. A. W. (1956). *An essay on the foundations of geometry*. New York: Dover.
88. Sedrakyan, N. (2010). Problem O174. *Mathematical Reflections, 5*, 4.
89. Sharygin, I. F. (1988). *Problems in plane geometry*. Imported Pubn.
90. Stanley, R. P. (2001). *Enumerative combinatorics* (Vols. 1 and 2, new ed.). Cambridge: Cambridge University Press.
91. Szekely, G. J. (Ed.) (1996). *Contests in higher mathematics. Miklos Schweitzer competitions 1962–1991*. New York: Springer.
92. Tao, T. (2006). *Solving mathematical problems. A personal perspective*. Oxford: Oxford University Press.
93. Tarski, A. (1951). *A decision method for elementary algebra and geometry*. Berkeley: University of California Press.
94. Tiel, J. V. (1984). *Convex analysis. An introductory text*. Chichester: Wiley.

95. Yaglom, I. M. (1962). *Geometric transformations* (Vol. I). Washington: Math. Assoc. of America.
96. Yaglom, I. M. (1968). *Geometric transformations* (Vol. II). Washington: Math. Assoc. of America.
97. Yaglom, I. M. (1973). *Geometric transformations* (Vol. III). Washington: Math. Assoc. of America.
98. Wells, D. (1991). *The Penguin dictionary of curious and interesting geometry*. London: Penguin.
99. Wiles, A. (1995). Modular elliptic curves and Fermat's last theorem. *Annals of Mathematics*, *141*(3), 443–551.

Index of Symbols

$\mathbb{N}$: The set of natural numbers $1, 2, 3, \ldots, n, \ldots$
$\mathbb{Z}$: The set of integers
$\mathbb{Z}^+$: The set of nonnegative integers
$\mathbb{Z}^-$: The set of nonpositive integers
$\mathbb{Z}^*$: The set of nonzero integers
$\mathbb{Q}$: The set of rational numbers
$\mathbb{Q}^+$: The set of nonnegative rational numbers
$\mathbb{Q}^-$: The set of nonpositive rational numbers
$\mathbb{R}$: The set of real numbers
$\mathbb{R}^+$: The set of nonnegative real numbers
$\mathbb{R}^-$: The set of nonpositive real numbers
$\mathbb{C}$: The set of complex numbers
$\mathbb{R}^2 = \mathbb{R} \times \mathbb{R} = \{(x, y) : x, y \in \mathbb{R}\}$
$\mathbb{R}^3 = \mathbb{R} \times \mathbb{R} \times \mathbb{R} = \{(x, y, z) : x, y, z \in \mathbb{R}\}$
E^2: The Euclidean plane
E^3: The Euclidean 3-dimensional space
π: Ratio of the circumference of circle to diameter, $\pi \cong 3.14159265358\ldots$
e: Base of natural logarithm, $e \cong 2.718281828459\ldots$
$a \in A$: a is an element of the set A
$a \notin A$: a is not an element of the set A
$A \cup B$: Union of two sets A, B
$A \cap B$: Intersection of two sets A, B
$A \times B$: Cartesian product of two sets A, B
$a \Rightarrow b$: if a then b
$a \Leftarrow b$: if b then a
$a \Leftrightarrow b$: a if and only if b
$\emptyset$: Empty set
$A \subseteq B$: A is a subset of B
$n! = 1 \cdot 2 \cdot 3 \cdots n$, where $n \in \mathbb{N}$
$\widehat{ABC}$: Angle with sides $\overrightarrow{BA}$ and $\overrightarrow{BC}$
$l \perp m$: Line l perpendicular to line m

S.E. Louridas, M.Th. Rassias, *Problem-Solving and Selected Topics in Euclidean Geometry*, DOI 10.1007/978-1-4614-7273-5,

$\overline{AB} \perp \overline{CF}$: Segment $\overline{AB}$ is perpendicular to segment $\overline{CF}$
$\triangle ABC$: Triangle ABC
$\widehat{ABC} \cong \widehat{XYZ}$: $\widehat{ABC}$ is congruent to $\widehat{XYZ}$
$\triangle ABC \cong \triangle XYZ$: $\triangle ABC$ is congruent to $\triangle XYZ$
$\triangle ABC \sim \triangle XYZ$: $\triangle ABC$ is similar to $\triangle XYZ$
$l \parallel m$: Line l is parallel to line m
$\overline{AB} \parallel \overline{CD}$: Segment $\overline{AB}$ is parallel to segment $\overline{CD}$
$S_1 = \mathrm{Inv}_{(O,\lambda)}\, S_2$: S_1 is the inverse shape of the shape S_2 with respect to the pole O and power λ
$\square$: End of the solution or the proof

Subject Index

S.E. Louridas, M.Th. Rassias, *Problem-Solving and Selected Topics in Euclidean Geometry*, DOI 10.1007/978-1-4614-7273-5,

Printed in the United States
By Bookmasters